Memoirs of the American Mathematical Society

Number 203

F. A. Howes

Boundary-interior layer interactions in nonlinear singular perturbation theory

Published by the

AMERICAN MATHEMATICAL SOCIETY

Providence, Rhode Island

VOLUME 15 · NUMBER 203 (first of two numbers) · JULY 1978

Abstract

For three classes of singularly perturbed boundary value problems we study the existence of solutions which possess boundary, shock and corner layer behavior and we examine how these nonuniformities arise and how they influence one another. The keys to our analysis are the stability properties of solutions of corresponding reduced problems and the geometric properties of solutions of the boundary value problems inside such layers. Several examples of the theory are discussed in detail with a view to illustrating the naturalness of our approach.

AMS(MOS) Subject Classifications (1970). Primary 34E15; Secondary 34B15.

Library of Congress Cataloging in Publication Data CIP

Howes, Frederick A 1948-
Boundary-interior layer interactions in nonlinear singular perturbation theory.

(Memoirs of the American Mathematical Society ; no. 203)
"Volume 15 ... first of two numbers."
"CODEN: MAMCAU."
Bibliography: p.
1. Boundary value problems--Numerical solutions.
2. Perturbation (Mathematics). I. Title. II. Series: American Mathematical Society. Memoirs ; no. 203.
QA3.A57 no. 203 [QA379] 510'.8s [515'.35] 78-8693
ISBN 0-8218-2203-9

TABLE OF CONTENTS

Section 1. Introduction. 1

Section 2. Stability of Reduced Solutions. 8

Section 3. Geometric Character of Boundary and Shock Layer Behavior. 12

Section 4. The Problem (P_1). 20

Section 5. The Problem (P_2). 35

Section 6. The Problem (P_3). 49

Section 7. Examples of (P_1). 60

Section 8. Examples of (P_2). 79

Section 9. Examples of (P_3). 89

References 106

ACKNOWLEDGMENTS

It is my pleasure to thank Mrs. Lynne Moberg and Mrs. Shirley Ward for typing the original manuscript and Ms. Kathy Swedell for typing the final copy. I also want to express my gratitude to the National Science Foundation for its generous support of my research into singular perturbation theory over the past few years. Finally I am deeply indebted to Bob O'Malley and Wolfgang Wasow for reading the original manuscript and suggesting many improvements.

To a Beautiful Princess

(whose feet will always be too little)

BOUNDARY-INTERIOR LAYER INTERACTIONS

IN

NONLINEAR SINGULAR PERTURBATION THEORY

1. Introduction. We consider here the existence and the asymptotic behavior of solutions of the scalar boundary value problem

$$\text{(P)}\qquad \begin{aligned} &\epsilon y'' = F(t,y,y') , \quad a < t < b , \\ &y(a,\epsilon) = A , \; y(b,\epsilon) = B , \end{aligned}$$

for small positive values of the parameter ϵ when F is at most a quadratic function of y'. More specifically we are interested in demonstrating how nonuniformities in the solutions of (P) (as functions of t and ϵ) arise and how nonuniform behavior at the boundaries (that is, boundary layer behavior) is related to nonuniform behavior in (a,b) (that is, interior layer behavior). The discussion divides naturally into a consideration of three classes of functions F; namely,

(P_1) $\quad F(t,y,y') = h(t,y)$,

(P_2) $\quad F(t,y,y') = f(t,y)y' + g(t,y)$

and

(P_3) $\quad F(t,y,y') = p(t,y)y'^2 + q(t,y)y' + r(t,y)$.

Received by the editors April 7, 1977, and in revised form February 23, 1978.
Supported in part by the National Science Foundation under Grant no. MCS 76-05979 .

It is precisely for these types of problems that the boundary and interior layer behavior described in this paper are possible (cf. [34] and [32; Chap. 2]) .

Before briefly reviewing some of the literature on such problems we outline here by means of several simple examples the kinds of asymptotic phenomena which will be our main concern throughout this paper. The first example is the linear problem

$$(\Pi_1) \qquad \begin{aligned} &\varepsilon y'' = my \ , \quad 0 < t < 1 \ , \\ &y(0,\varepsilon) = 1 \ , \quad y(1,\varepsilon) = 2 \end{aligned}$$

where m is a positive constant and $\varepsilon > 0$ is a small parameter. A direct calculation shows that the solution of (Π_1) is given (to asymptotically zero terms) by

$$y(t,\varepsilon) = \exp[-(m\varepsilon^{-1})^{1/2} t] + 2\exp[-(m\varepsilon^{-1})^{1/2}(1-t)]$$

and therefore $\lim_{\varepsilon\to 0^+} y(t,\varepsilon) = 0$ for $0 < t < 1$. However,

$$\lim_{t\to 0^+}\lim_{\varepsilon\to 0^+} y(t,\varepsilon) = 0 \neq 1 = \lim_{\varepsilon\to 0^+}\lim_{t\to 0^+} y(t,\varepsilon)$$

and

$$\lim_{t\to 1^-}\lim_{\varepsilon\to 0^+} y(t,\varepsilon) = 0 \neq 2 = \lim_{\varepsilon\to 0^+}\lim_{t\to 1^-} y(t,\varepsilon) \ ,$$

that is, $y(t,\varepsilon)$ exhibits boundary layer behavior at both endpoints; see figure 1. We note also

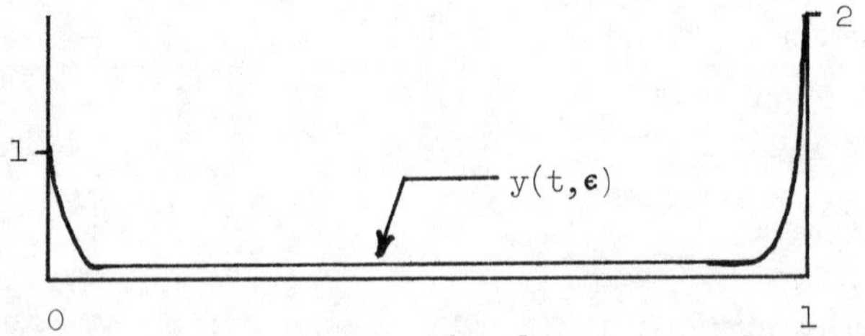

Fig. 1

The Solution of (Π_1)

that the limit of y within $(0,1)$, namely $u \equiv 0$, is a solution of the reduced equation $mu = 0$ obtained from (Π_1) by setting $\epsilon = 0$. The restriction that m be positive is essential for the solution of (Π_1) to exist for all small values of $\epsilon > 0$ and to converge to 0 in $(0,1)$. In this case it is natural to call $u \equiv 0$ a stable solution of the reduced equation.

The second example is a nonlinear problem which illustrates another type of nonuniformity that solutions of (P) can possess; namely,

$$(\Pi_2) \qquad \begin{aligned} &\epsilon y'' = yy', \; -1 < t < 1, \\ &y(-1,\epsilon) = A, \; y(1,\epsilon) = B \end{aligned}$$

where $A = -B > 0$. As in the previous example the solution $y = y(t,\epsilon)$ of (Π_2) can be found by quadratures (cf. [37] and [29; Chap. 1]) and it follows that

$$y(t,\epsilon) = -A(1-\exp[-At\epsilon^{-1}])(1+\exp[-At\epsilon^{-1}])^{-1} .$$

Consequently, $\lim\limits_{\epsilon\to 0^+} y(t,\epsilon) = \begin{cases} A, & -1 \le t < 0, \\ B, & 0 < t \le 1, \end{cases}$

and since $y(0,\epsilon)$ is asymptotically zero,

$$\lim_{t\to 0^-} \lim_{\epsilon\to 0^+} y(t,\epsilon) = A \neq 0 = \lim_{\epsilon\to 0^+} \lim_{t\to 0^-} y(t,\epsilon)$$

and

$$\lim_{t\to 0^+} \lim_{\epsilon\to 0^+} y(t,\epsilon) = B \neq 0 = \lim_{\epsilon\to 0^+} \lim_{t\to 0^+} y(t,\epsilon) .$$

Thus $y(t,\epsilon)$ behaves nonuniformly at $t = 0$ and we will say that there is a shock layer (or an interior transition layer) at the origin; see figure 2 .

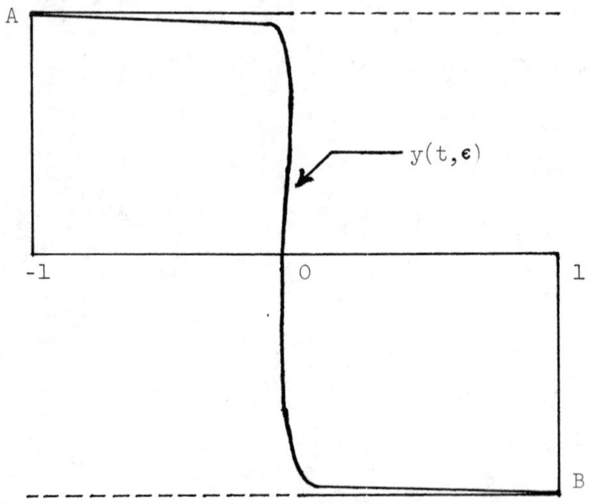

Fig. 2

The Solution of (π_2) .

The constant limiting values $u_L \equiv A$ and $u_R \equiv B$ of y on $[-1,0)$ and $(0,1]$, respectively, are clearly solutions of the reduced equation $uu' = 0$. We note that (cf. [37] and [29; Chap. 1]) it is only for this (very special) choice of boundary conditions that the solution of (π_2) exhibits the shock layer behavior depicted in figure 2 .

Our final introductory example concerns the case in which a solution of (P) behaves uniformly throughout $[a,b]$ but has a derivative which behaves nonuniformly at a point in (a,b); namely,

$$\epsilon y'' = 1 - y'^2 \ , \quad 0 < t < 1 \ ,$$

(π_3)

$$y(0,\epsilon) = A \ , \ y(1,\epsilon) = B$$

where $0 \leq |A-B| < 1$. Once more this problem can be solved by quadratures (cf. [14] or [37]) and the solution (to terms of order $O(\epsilon)$)

$$y(t,\epsilon) = \epsilon \ln \cosh[\epsilon^{-1}(t + \frac{B-1-A}{2})] + \frac{A+B-1}{2} \ .$$

From this we see that for $t_0 = \frac{1}{2}(A-B+1)$ (which belongs to $(0,1)$ by our restriction on A and B)

$$\lim_{\epsilon\to 0^+} y(t,\epsilon) = \begin{cases} A - t \ , & 0 \leq t \leq t_0 \ , \\ t + B - 1 \ , & t_0 \leq t \leq 1 \ , \end{cases}$$

and

$$\lim_{\epsilon\to 0^+} y'(t,\epsilon) = \begin{cases} -1 \ , & 0 \leq t < t_0 \ , \\ 1 \ , & t_0 < t \leq 1 \ . \end{cases}$$

Thus $y'(t,\epsilon)$ behaves nonuniformly at $t = t_0$ which is the point of intersection of $u_L(t) = A - t$ and $u_R(t) = t + B - 1$ in $(0,1)$; see figure 3 .

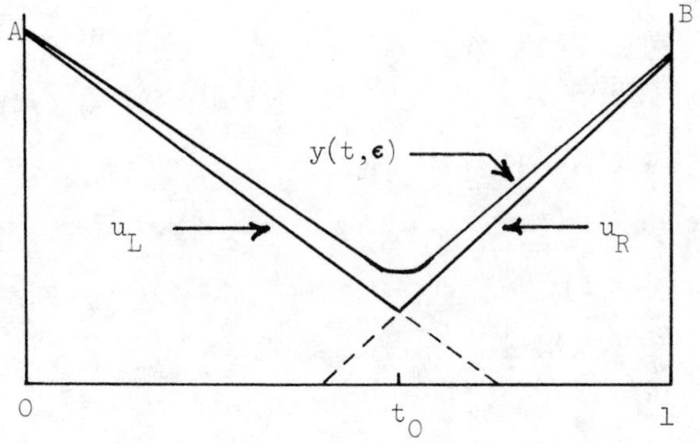

Fig. 3

The Solution of (π_3)

The functions u_L and u_R are solutions of the corresponding reduced equation $1 - u'^2 = 0$ which satisfy $u_L(0) = A$ and $u_R(1) = B$. We note that $\tilde{u}_L(t) = t + A$ and $\tilde{u}_R(t) = B + 1 - t$ also have these properties; however, the asymptotic behavior of the solution of (Π_3) is described by the pair (u_L, u_R) and not by $(\tilde{u}_L, \tilde{u}_R)$. In a sense to be made precise shortly the functions u_L and u_R are <u>stable</u> solutions of $1 - u'^2 = 0$ while $\tilde{u}_L$ and $\tilde{u}_R$ are unstable.

It may be helpful if the reader keeps these examples in mind at the outset of this paper since they serve to motivate much of the discussion in the first few sections. We turn now to a brief review of the literature on the asymptotic phenomena to be considered below.

The problems (P_1) and (P_2) have received a great deal of attention in the last twenty-five years; many basic results are summarized in the article of Vasil'eva [32] and in the books of Wasow [36; Chap. 10] and O'Malley [29]. Our interest in them has been motivated by recent work of Fife [9-12] on (P_1) and Flaherty and O'Malley [13] on (P_1) and (P_2). Fife was able to give rather general sufficient conditions for the existence of solutions of (P_1) which possess the boundary and shock layer behavior given below. Using some of Fife's results together with earlier work of O'Malley [30] Flaherty and O'Malley made a fairly complete study of conditions which guarantee boundary layer behavior for solutions of (P_1) and (P_2). Their discussion of (P_2) is reminiscent of a much earlier treatment by Coddington and Levinson [5]. The approach of these five authors is similar in that the occurrence of boundary layer behavior (and interior layer behavior in the case of [10-12]) is investigated by studying the properties of functionals of the form

$$J_1[t] = \int_{u_1(t)}^{u_2(t)} h(t,s)ds \quad \text{in the case of} \quad (P_1) \quad \text{and}$$

$$J_2[t] = \int_{\tilde{u}_1(t)}^{\tilde{u}_2(t)} f(t,s)ds \quad \text{in the case of} \quad (P_2) \quad .$$

Here u_1, u_2 are certain solutions of $h(t,u) = 0$ and $\tilde{u}_1, \tilde{u}_2$ are certain solutions of $f(t,u)u' + g(t,u) = 0$. In the sections that follow we will prove their basic results in a more general setting and study the ways in which boundary and interior layer behavior are related to each other. By so doing we hope to demonstrate the value of using such functionals to classify the asymptotic behavior of solutions of (P_1) and (P_2) .

Problems of the form (P_3) have only recently come under a close scrutiny. Motivated by the work of Dorr, Parter and Shampine [7; Sec. 5] and Haber and Levinson [14], the author in [16; 18-22] has investigated some of the boundary and interior layer phenomena which solutions can display. It turns out that here too the behavior of solutions can be measured by a functional of the form

$$J_3[t] = \int_{u_1(t)}^{u_2(t)} p(t,s)ds$$

where u_1 and u_2 are appropriate solutions of $p(t,u)u'^2 + q(t,u)u' + r(t,u) = 0$.

In the next two sections we examine in detail the stability and geometric properties of solutions of the corresponding reduced equations of (P_1), (P_2) and (P_3) which are essential to our discussion. Following this we prove some general results about solutions of these problems which have nonuniformities in $[a,b]$ and we show how the functionals J_1, J_2 and J_3 can be used to describe this behavior.

Our theory is then applied to several model problems of past and current interest.

To prove the results of this paper we will employ the method of differential inequalities as first outlined by Nagumo [27] (cf. [17] for an illustration of the use of this method in related singular perturbation problems). Nagumo's method (as presented in [27] or [25]) is restricted in its applicability to problems (P) in which F is at most a quadratic function of y' ; however, it is possible to study problems of this type whose righthand sides have cubic or higher nonlinearities in y' (cf. [23]). This is done by using what is known as a generalized Nagumo condition and the reader is referred to a paper of Heidel [15] for a complete discussion of this useful concept.

2. Stability of Reduced Solutions. We list here the definitions of stability of the various solutions of the reduced equations (i.e., reduced solutions) of (P_1), (P_2) and (P_3) . In one form or another these definitions have been used in practically every treatment of such problems; however, for completness we present them now in full generality.

They are nothing more than restrictions on the signs of the partial derivatives $F_{y'}$ and $\partial_y^j F$ evaluated along certain solutions of the reduced equation. Here $F = F(t,y,y')$ denotes the righthand side of (P_1), (P_2) or (P_3) .

Consider first (P_1) and let $u = u(t)$ be a $C^{(2)}$-solution of the reduced equation $h(t,u) = 0$ which exists on $[a,b]$. The function u is said to be $(I)_q$-stable $(q \geq 0)$ if the partial derivatives $\partial_y^j h(t,u)$ exist for $1 \leq j \leq 2q+1$ and satisfy for t in $[a,b]$:

$$\partial_y^{\,j} h(t,u(t)) \equiv 0 \quad \text{for} \quad 1 \le j \le 2q \ ;$$

$$\partial_y^{\,2q+1} h(t,u(t)) > 0 \ .$$

Note that $(I)_q$-stability is independent of the relative position of $u(a)$ and A, and $u(b)$ and B. Suppose next that u satisfies $u(a) < A$ and $u(b) < B$. Then u is said to be $(II)_n$-stable $(n \ge 2)$ if $u'' \ge 0$ and if the partial derivatives $\partial_y^{\,j} h(t,u)$ exist for $1 \le j \le n$ and satisfy for t in $[a,b]$:

$$\partial_y^{\,j} h(t,u(t)) \ge 0 \quad \text{for} \quad 1 \le j \le n-1 \ ;$$

$$\partial_y^{\,n} h(t,u(t)) > 0 \ .$$

Finally suppose that u satisfies $u(a) > A$ and $u(b) > B$. Then u is said to be $(III)_n$-stable $(n \ge 2)$ if $u'' \le 0$ and if the partial derivatives $\partial_y^{\,j} h(t,u)$ exist for $1 \le j \le n$ and satisfy for t in $[a,b]$:

$$\partial_y^{\,j_0} h(t,u(t)) \ge 0 \ , \quad \partial_y^{\,j_e} h(t,u(t)) \le 0 \quad \text{for} \quad 1 \le j_0, j_e \le n-1$$

(where $j_0 (j_e)$ denotes an odd (even) integer); and

$$\partial_y^{\,n} h(t,u(t)) > 0 \quad \text{if} \quad n \quad \text{is odd} \ ,$$

$$\partial_y^{\,n} h(t,u(t)) < 0 \quad \text{if} \quad n \quad \text{is even} \ .$$

Such conditions were first introduced by Boglaev [1] ; more recent applications are given in [16], [18] and [22] . We note that in [9-13] the reduced solutions of (P_1) are assumed to be $(I)_0$-stable, that is, $h_y(t,u(t)) > 0$ for $a \le t \le b$. These definitions of stability can be motivated by considering simple nonlinear functions like $h(t,y) = y^{2q+1}$ or $h(t,y) = \pm y^{2n}$.

Consider next the quasilinear problem (P_2) . Here there are three principal types of reduced solutions, each with its own form of stability. Suppose first that the reduced equation $f(t,u)u' + g(t,u) = 0$ has a $C^{(2)}$-solution $u = u_L(t)$ such that $u_L(a) = A$. (Let us agree to denote any reduced solution which satisfies the lefthand (righthand) boundary condition generically by $u_L(u_R)$.) Then u_L is said to be stable on $[a,t_L]$ if $f(t,u_L(t)) > 0$ for $a \leq t < t_L \leq b$ and $f(t_L,u_L(t_L)) \geq 0$. Analogously suppose that this equation has a solution $u = u_R(t)$ (that is, $u_R(b) = B$) . Then u_R is said to be stable on $[t_R,b]$ if $f(t,u_R(t)) < 0$ for $a \leq t_R < t \leq b$ and $f(t_R,u_R(t_R)) \leq 0$. If $t_L = b$ $(t_R = a)$ we say that u_L (u_R) is globally stable, while if $t_L > t_R$ $(t_L < t_R)$ we say that the domains of stability of u_L and u_R overlap (do not overlap).

Suppose next that $u = u(t)$ is a solution of $f(t,u)u' + g(t,u) = 0$ and that u satisfies $u(a) \neq A$ $(u(b) \neq B)$. Then u is said to be stable with respect to boundary layer behavior at $t = a$ $(t = b)$ if

$$f(t,u(t)) < 0 \quad \text{for} \quad a \leq t \leq a+\delta$$

$$(f(t,u(t)) > 0 \quad \text{for} \quad b-\delta \leq t \leq b)$$

for some positive constant δ . This type of stability is discussed more fully in [21] and [22] . Finally suppose that the reduced equation has a solution $u = u_s(t)$ which satisfies $f(t,u_s(t)) \equiv 0$ for $a \leq t \leq b$. (Such a solution is called singular and is denoted generically by u_s .) A solution u_s is said to be stable if it is $(I)_q$-, $(II)_n$- , or $(III)_n$-stable in the sense described above with $h(t,y)$ replaced by $f(t,y)\,u_s'(t) + g(t,y)$. For example, u_s is $(I)_0$-stable if

$$f_y(t,u_s(t))u_s'(t) + g_y(t,u_s(t)) > 0 \quad \text{for} \quad a \leq t \leq b \ .$$

It is possible to motivate the definitions of (u_L,u_R)-stability by considering the simple function $F(t,y,y') = \pm\, ky'$, $k > 0$.

Finally consider the problem (P_3) . Several definitions of stability have been given in [21] and [22] ; however, here we only require the following types. Suppose first that the reduced equation $p(t,u)u'^2 + q(t,u)u' + r(t,u) = 0$ has a $C^{(2)}$-solution $u = u_L(t)$ (that is, $u_L(a) = A$) then u_L is said to be stable on $[a,t_L]$ if $\varphi(t,u_L(t)) = 2p(t,u_L(t))\,u_L'(t) + q(t,u_L(t)) > 0$ for $a \leq t < t_L \leq b$ and $\varphi(t_L,u_L(t_L)) \geq 0$. Similarly a reduced solution $u = u_R(t)$ is said to be stable on $[t_R,b]$ if $\varphi(t,u_R(t)) < 0$ for $a \leq t_R < t \leq b$ and $\varphi(t_R,u_R(t_R)) \leq 0$. We also want to define the concept of boundary layer stability for reduced solutions; namely, suppose that there is a solution $u = u(t)$ such that $u(a) \neq A$ $(u(b) \neq B)$. Then u is said to be stable with respect to boundary layer behavior at $t = a (t = b)$ if $\varphi(t,u(t)) < 0$ for $a \leq t \leq a+\delta$ $(\varphi(t,u(t)) > 0$ for $b-\delta \leq t \leq b)$ for a positive constant δ . Finally, reduced equations of the form $p(t,u)u'^2 + q(t,u)u' + r(t,u) = 0$ often have singular solutions $u = u_s(t)$ (that is, $p(t,u_s(t)) = q(t,u_s(t)) = r(t,u_s(t)) \equiv 0$) . In this case u_s is said to be stable if it is $(I)_q$-, $(II)_n$-, or $(III)_n$-stable in the sense described above with $h(t,y)$ replaced by $p(t,y)u_s'(t)^2 + q(t,y)u_s'(t) + r(t,y)$.

The reader can easily verify that the limiting behavior of the solutions of the three introductory examples (Π_1) - (Π_3) is described by solutions of appropriate reduced problems which are stable in the senses just described. More generally we will see below that in order to characterize the limiting behavior of solutions of the problems (P_1), (P_2) and (P_3) we can focus our attention almost exclusively on stable solutions of the reduced equations.

3. Geometric Character of Boundary and Shock Layer Behavior.

We now examine the convexity properties which solutions of (P_1), (P_2) and (P_3) possess inside of boundary and shock layers. The facts noted here are obvious but they do not seem to have been stated explicitly before in the literature. The following simple notions will prove of great use in enabling us to ascribe the correct behavior to solutions of these problems.

Consider first

$$(P_1) \qquad \begin{aligned} &\varepsilon y'' = h(t,y), \quad a < t < b, \\ &y(a,\varepsilon) = A, \quad y(b,\varepsilon) = B. \end{aligned}$$

Suppose that $h(t,u) = 0$ has a stable solution $u = u(t)$ on $[a,b]$ such that $u(a) < A$ (and for simplicity, $u(b) = B$). If (P_1) has a solution $y = y(t,\varepsilon)$ such that

$$\lim_{\varepsilon\to 0^+} y(t,\varepsilon) = u(t) \quad \text{for} \quad a < t \le b$$

then there is a boundary layer near $t = a$ within which y rises sharply to satisfy the boundary condition $y(a,\varepsilon) = A$. (Hereafter when such a situation obtains we shall say that u supports a boundary layer at $t = a$. Similarly if $\lim_{\epsilon\to 0+} y(t,\epsilon) = u(t)$ for $a \le t < b$ we shall say that u supports a boundary layer at $t = b$.) Clearly $h(a,\xi)$ must be positive for $\xi > u(a)$ and $(\xi-u(a))$ sufficiently small since $\epsilon y''(t,\epsilon) \simeq h(a,\xi)$ for t near a, and so y is convex. If $h(a,\xi) > 0$ for all ξ in $(u(a),A]$ then y is convex throughout the boundary layer; see figure 4.

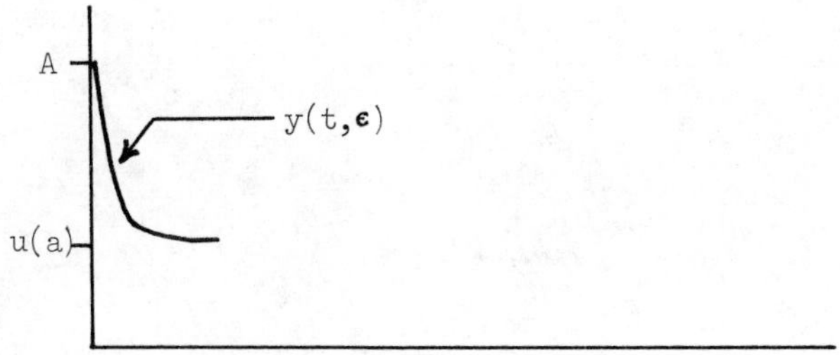

Fig. 4

Convex (U-) Boundary Layer at $t = a$.

However if there is a point ξ^* such that $h(a,\xi) > 0$ for $u(a) < \xi < \xi^*$, $h(a,\xi^*) = 0$ and $h(a,\xi) < 0$ for $\xi^* < \xi \le A$, then y is convex near the outer edge of the layer but concave at $t = a$; see figure 5.

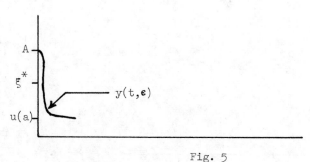

Fig. 5

Convex-Concave (Z-) Boundary Layer at $t = a$.

What this means is that there is another solution of $h(t,u) = 0$ lying between $u(t)$ and the horizontal line $y \equiv A$, at least near $t = a$. Finally the boundary layer structure can be even more complicated in that y can be alternately concave and convex within the layer. An example of this is discussed in Section 7. In the first case we shall say that $y(t,\epsilon)$ has a convex (that is, U-) boundary layer at $t = a$, while in the second we shall say that $y(t,\epsilon)$ has a convex-concave (that is, Z-) boundary layer there. Similarly if $u(a) > A$ and if $\lim_{\epsilon\to 0+} y(t,\epsilon) = u(t)$ for $a < t \le b$ then $y(t,\epsilon)$ can be concave inside of the layer (that is, $y'' < 0$) or concave at the outer edge of the layer becoming convex at $t = a$; see figures 6 and 7.

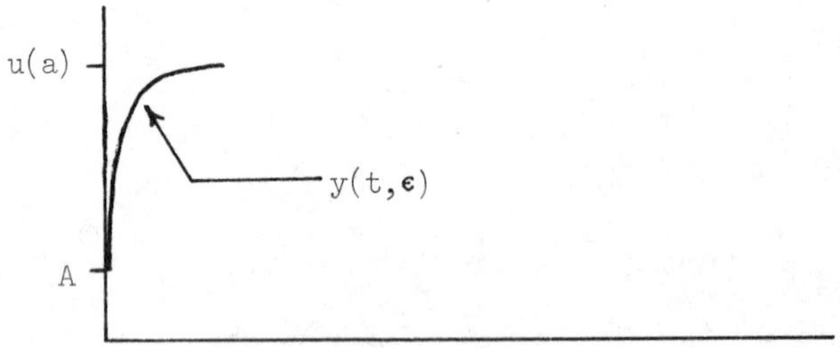

Fig. 6

Concave (∩-) Boundary Layer at $t = a$.

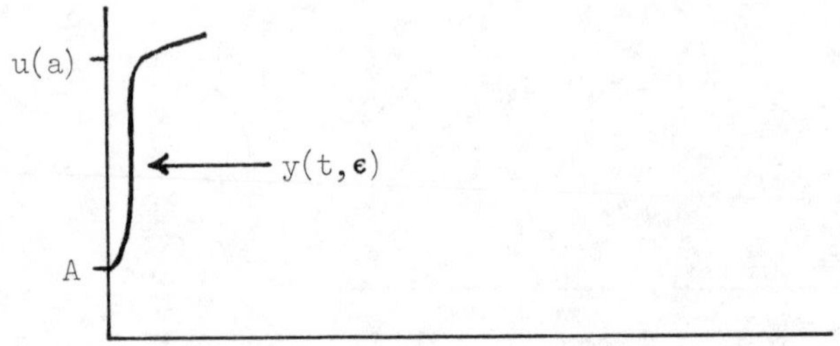

Fig. 7

Concave-Convex (S-) Boundary Layer at $t = a$.

Here too more complicated behavior is possible. In the first case we shall say that y has a concave (that is, $\cap$-) boundary layer at $t = a$, while in the second we shall say that y has a concave-convex (that is, S-) boundary layer there. Note that in the case of an S-layer $h(a,\xi)$ is zero for some ξ^* in $(A,u(a))$.

Suppose next that $h(t,u) = 0$ has three solutions u_1, $\tilde{u}$ and u_2 (and for simplicity, $u_1(a) = A$, $u_2(b) = B$). In addition, suppose that u_1 and u_2 are stable, while $\tilde{u}$ is unstable (for example, $h_y(t,\tilde{u}) < 0$). Then if (P_1) has a solution $y = y(t,\epsilon)$ such that

$$\lim_{\epsilon\to 0^+} y(t,\epsilon) = \begin{cases} u_1(t), & a \leq t < t^*, \\ u_2(t), & t^* < t \leq b, \end{cases}$$

we say that there is a shock layer at $t = t^*$ since y moves rapidly from a neighborhood of $u_1(t)$ to one of $u_2(t)$ as t crosses t^*; see figure 8. (Hereafter when such a situation obtains we shall say that u_1 and u_2 support a shock layer at $t = t^*$.) Note that the unstable solution $\tilde{u}$ is not involved in the limiting solution.

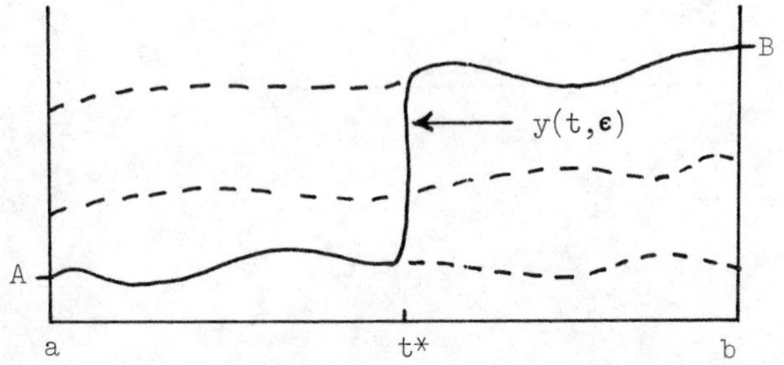

Fig. 8

S-Shock Layer at $t = t^*$.

Clearly $h(t^*+0,\eta)$ must be negative for $\eta < u_2(t^*+0)$ and $(u_2(t^*+0) - \eta)$ small, while $h(t^*-0,\eta)$ must be positive for $\eta > u_1(t^*-0)$ and $(\eta - u_1(t^*-0))$ small. Moreover $h(\hat{t},\hat{\eta}) = 0$ for some $\hat{t}$ near t^* and $\hat{\eta}$ in $(u_1(t^*-0), u_2(t^*+0))$. We shall say that in such a case y has a convex-concave (that is, S-) shock layer at t^*. If on the other hand $u_1 > \tilde{u} > u_2$ and if

$$\lim_{\epsilon\to 0^+} y(t,\epsilon) = \begin{cases} u_1(t), & a \le t < t^*, \\ u_2(t), & t^* < t \le b, \end{cases}$$

then we shall say that y has a concave-convex (that is, Z-) shock layer at t^*; see figure 9.

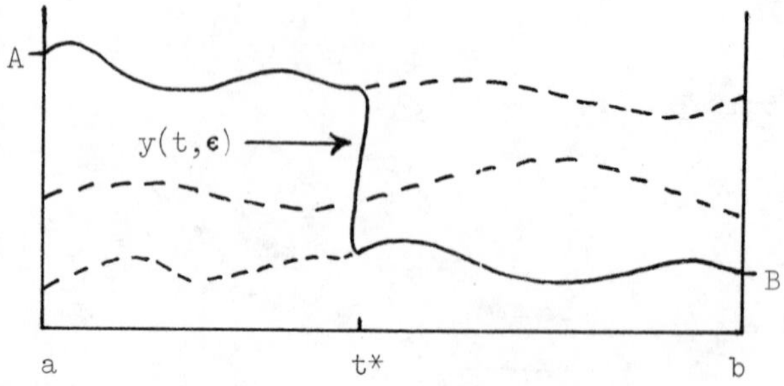

Fig. 9

Z-Shock Layer at $t = t^*$.

Consider next the quasilinear problem

$$\epsilon y'' = f(t,y)y' + g(t,y), \quad a < t < b ,$$

(P_2)

$$y(a,\epsilon) = A \ , \ y(b,\epsilon) = B \ .$$

Suppose first that (P_2) has a solution $y = y(t,\epsilon)$ with a boundary layer at $t = a$, that is,

$$\lim_{\epsilon \to 0^+} y(t,\epsilon) = u_R(t) \quad \text{for} \quad a < t \leq b$$

where u_R is a stable solution of $f(t,u)u' + g(t,u) = 0$ satisfying $u_R(b) = B$. Since $y'(t,\epsilon) = O(\epsilon^{-1})$ inside of such a layer (cf. [34], [32; Chap. 2]) the convexity of any solution near $t = a$ is determined by discarding the g-term in (P_2) and looking at (*) $\epsilon y'' = f(t,y)y'$. The stability of u_R requires that $f(t,u_R(t)) < 0$, and therefore $f(a,\eta) < 0$ for η near $u_R(a)$. Assume that $u_R(a)$ is greater than A, then y has to be concave near the outer edge of the boundary layer which is consistent with (*) since $y'(t,\epsilon) > 0$ for t near a. If $f(a,\eta) < 0$ for all η in $[A,u_R(a))$ then y is concave throughout the layer since $\epsilon y'' \simeq f(a,\eta)y' < 0$. However if there is a point η^* such that $f(a,\eta) < 0$ for $\eta^* < \eta < u_R(a)$, $f(a,\eta^*) = 0$, and $f(a,\eta) > 0$ for $A \leq \eta < \eta^*$, then y is clearly concave at the outer edge of the layer but convex at $t = a$; cf. figures 6 and 7. As in the case of (P_1) we shall say that y has a concave (that is, $\cap$-) boundary layer at $t = a$ if $f(a,\eta) < 0$ for $A \leq \eta < u_R(a)$ and that y has a concave-convex (that is, S-) layer there if $f(a,\eta^*) = 0$ for some η^* in $(A,u_R(a))$. Similarly if $u_R(a) < A$ then y can have a convex (that is, $\cup$-) layer at $t = a$ if $f(a,\eta) < 0$ for $u_R(a) < \eta \leq A$ or a convex-concave (that is, Z-) layer there if $f(a,\eta^*) = 0$ at a point in η^* in $(u_R(a),A)$. It is clear that if y has either an S-layer or a Z-layer at $t = a$ then $u_R(t)$ and $y \equiv A$ must be separated by a singular reduced solution, at least near $t = a$.

Suppose next that (P_2) has a solution $y = y(t,\epsilon)$ with an S-shock layer at $t = t^*$ in (a,b), that is,

$$\lim_{\epsilon \to 0^+} y(t,\epsilon) = \begin{cases} u_L(t), & a \leq t < t^*, \\ u_R(t), & t^* < t \leq b. \end{cases}$$

Here $u_L < u_R$ are stable reduced solutions which are separated by a singular solution u_s. Such a shock layer structure is consistent with the equation (*) since

$$\epsilon y''(t^*-0,\epsilon) \simeq f(t^*-0,\eta)y'(t^*-0,\epsilon) > 0$$

if $(\eta-u_L(t^*-0))$ is small and positive (since u_L is stable and $y'(t,\epsilon) > 0$ for t near t^*) and

$$\epsilon y''(t^*+0,\epsilon) \simeq f(t^*+0,\eta)\, y'(t^*+0,\epsilon) < 0$$

if $(u_R(t^*+0) - \eta)$ is small and positive (since u_R is stable and $y'(t,\epsilon) > 0$ for t near t^*). In addition, $f(\hat{t},u_s(\hat{t})) = g(\hat{t},u_s(\hat{t})) = 0$ for a $\hat{t}$ near t^* which implies that $\epsilon y''(\hat{t},\epsilon) = 0$. This is of course consistent with the geometry of the S-layer. Analogous remarks can also be given concerning the case of a Z-shock layer in the case that $u_L > u_s > u_R$.

Consider finally the problem

$$(P_3)\qquad \begin{aligned} &\epsilon y'' = p(t,y)y'^2 + q(t,y)y' + r(t,y) \ , \qquad a < t < b \ , \\ &y(a,\epsilon) = A \ , \quad y(b,\epsilon) = B \ , \end{aligned}$$

and suppose first that there is a solution $y = y(t,\epsilon)$ of (P_3) which has boundary layer behavior at $t = b$, that is,

$$\lim_{\epsilon\to 0^+} y(t,\epsilon) = u_L(t) \quad \text{for} \quad a \leq t < b \ .$$

Here u_L is a stable reduced solution. If $u_L(b) < B$ then inside of the boundary layer $y'(t,\epsilon) > 0$ and $y'(t,\epsilon) = O\ (\exp[k\epsilon^{-1}])$ for some $k > 0$ (cf. [34] and [32; Chapt. 2]). Consequently the curvature of y near $t = b$ is determined by the equation

$$(**)\qquad \epsilon y'' \simeq p(t,y)\, y'^2$$

and since y is convex at the outer edge of the layer it is necessary that $p(b,\eta) > 0$ for $(\eta - u_L(b))$ positive and small. If $p(b,\eta) > 0$ for $u_L(b) < \eta \leq B$ then clearly y is convex throughout the layer and in this case we shall say as before that y has a U-boundary layer at $t = b$. However if there is an η^* suth that $p(b,\eta) > 0$ for $u_L(b) < \eta < \eta^*$, $p(b,\eta^*) = 0$ and $p(b,\eta) < 0$ for $\eta^* < \eta \leq B$, it follows that y is convex then concave inside the layer. We shall say that in this case y has an S-boundary layer at $t = b$. Similarly if $u_L(b) > B$ it is necessary that $p(b,\eta) < 0$ for $(u_L(b)-\eta)$ small and positive. The solution y has a ∩-boundary layer or a

Z-boundary layer depending on whether $p(b,\eta) < 0$ for $B \leq \eta < u_L(b)$ or $p(b,\eta^*) = 0$ at a point η^* between B and $u_L(b)$.

Suppose finally that a solution $y = y(t,\epsilon)$ of (P_3) has an S-shock layer at $t = t^*$ in (a,b) , that is,

$$\lim_{\epsilon \to 0^+} y(t,\epsilon) = \begin{cases} u_L(t) , & a \leq t < t^* , \\ u_R(t), & t^* < t \leq b , \end{cases}$$

where $u_L < u_s < u_R$ for stable reduced solutions u_L, u_R and a singular solution u_s . It is clearly necessary that $p(t^*+0,\eta) < 0$ for $(u_R(t^*+0)-\eta)$ small and positive and $p(t^*-0,\eta) > 0$ for $(\eta-u_L(t^*-0))$ small and positive, as follows from (**) . Similarly if $u_L > u_s > u_R$ then the inequalities

$$p(t^*-0,\eta) < 0 \quad \text{for} \quad (u_L(t^*-0)-\eta) \quad \text{small and positive}$$

and

$$p(t^*+0,\eta) > 0 \quad \text{for} \quad (\eta-u_R(t^*+0)) \quad \text{small and positive}$$

are necessary for y to have a Z-shock layer at t^* .

In the following sections we will give precise necessary and sufficient conditions (via the functionals J_1, J_2 and J_3) for the existence of these various layer phenomena. The geometric reasoning presented in this section is however useful for visualizing and motivating such conditions.

4. The Problem (P_1). We study now the behavior of solutions of

$$(P_1) \qquad \begin{aligned} &\epsilon y'' = h(t,y) , \qquad a < t < b , \\ &y(a,\epsilon) = A , \qquad y(b,\epsilon) = B . \end{aligned}$$

The theorems presented below are generalizations of some results of Fife [9-12].

Suppose that the reduced equation of (P_1) has a stable solution $u = u(t)$. The first group of theorems gives sufficient conditions that (P_1) have a solution $y = y(t,\epsilon)$ such that

$$\lim_{\epsilon\to 0^+} y(t,\epsilon) = u(t) \quad \text{for} \quad a < t < b \ .$$

Theorem 4.1. Assume that

(1) the equation $h(t,u) = 0$ has a solution $u = u(t)$ of class $C^{(2)}[a,b]$;

(2) the function h is of class $C^{(1)}$ with respect to t and of class $C^{(2q+1)} (q \geq 0)$ with respect to y in $\mathcal{D}$: $a \leq t \leq b$, $y = u(t) + d(t,\epsilon)$ where $d = O(\epsilon^{1/(2q+1)})$ for $a+\delta \leq t \leq b-\delta$, $d \geq |A - u(a)|$ for $a \leq t < a+\delta$ and $d \geq |B - u(b)|$ for $b-\delta < t \leq b$ with δ a small positive constant;

(3) the function u is $(I)_q$-stable on $[a,b]$;

(4)

$$\int_{u(a)}^{\eta} h(a,s)ds > 0 \qquad \begin{array}{l} \text{for } u(a) < \eta \leq A \text{ if } u(a) < A \ , \\ \text{for } A \leq \eta < u(a) \text{ if } u(a) > A \ , \end{array}$$

and

$$\int_{u(b)}^{\eta} h(b,s)ds > 0 \qquad \begin{array}{l} \text{for } u(b) < \eta \leq B \text{ if } u(b) < B \ , \\ \text{for } B \leq \eta < u(b) \text{ if } u(b) > B \ . \end{array}$$

Then (P_1) has a solution $y = y(t,\epsilon)$ for each sufficiently small $\epsilon > 0$. In addition, for $a \leq t \leq b$

$$y(t,\epsilon) = u(t) + w_L(t,\epsilon) + w_R(t,\epsilon) + O(\epsilon^{1/(2q+1)})$$

<u>where</u> $w_L(t,\epsilon)$ $(w_R(t,\epsilon))$ <u>is a</u> $O(\sqrt{\epsilon})$-<u>boundary layer function at</u> $t = a$ $(t = b)$. (<u>That is,</u> $w_L(t,\epsilon) = O(|A-u(a)|)$ <u>for</u> $a \le t \le a+c_1\sqrt{\epsilon}$, $w_R(t,\epsilon) = O(|B-u(b)|)$ <u>for</u> $b - c_2\sqrt{\epsilon} \le t \le b$ <u>and</u> $\lim_{\epsilon\to 0+} w_{L,R}(t,\epsilon) = 0$ <u>for each fixed</u> t <u>in</u> (a,b).)

<u>Proof</u>. Suppose for simplicity that $u(a) > A$ and $u(b) = B$. By a classical result of Nagumo [27] (see also [25])to prove this result it is enough to construct functions α,β of class $C^{(2)}[a,b]$ such that $\alpha \le \beta$, $\alpha(a,\epsilon) \le A \le \beta(a,\epsilon)$, $\alpha(b,\epsilon) \le B \le \beta(b,\epsilon)$ and

$$\epsilon\alpha'' \ge h(t,\alpha) \;,\quad \epsilon\beta'' \le h(t,\beta) \quad \text{for} \quad a < t < b \;.$$

Define then for $\epsilon > 0$, $a \le t \le b$ and $\epsilon_1 = (\epsilon\gamma m^{-1})^{1/(2q+1)}$,

$$\alpha(t,\epsilon) = u(t) + w_L(t,\epsilon) - \epsilon_1$$

and

$$\beta(t,\epsilon) = u(t) + \epsilon_1$$

where $w_L < 0$ is a certain $C^{(2)}$-solution of the boundary layer inequality (*) $\epsilon\tilde{y}'' > h(a,\tilde{y}+u(a))$ satisfying $w_L(a,\epsilon) = A - u(a)$ and $\lim_{\epsilon\to 0^+} w_L(t,\epsilon) = 0$ for each fixed $t > a$. In addition, m is a positive constant such that $\partial_y^{2q+1} h(t,u) > m$ and γ is a positive constant to be chosen later. The existence of m is guaranteed by assumption (3). Clearly we have that $\alpha \le \beta$, $\alpha(a,\epsilon) \le A \le \beta(a,\epsilon)$ and $\alpha(b,\epsilon) \le B \le \beta(b,\epsilon)$, and just as easily for $a < t < b$

$$h(t,\beta) - \epsilon\beta'' = \sum_{j=0}^{2q} \frac{1}{j!} \partial_y^{\,j} h(t,u)\epsilon_1^{\,j}$$

$$+ \frac{1}{(2q+1)!} \partial_y^{2q+1} h(t,u+\theta\epsilon_1)\epsilon_1^{\,2q+1} - \epsilon u''$$

$$\geq \frac{\epsilon\gamma}{(2q+1)!} - \epsilon M \text{ for } \epsilon \text{ sufficiently small },$$

where $0 < \theta < 1$ and $|u''(t)| \leq M$. Thus for $\gamma \geq (2q+1)!M$ $h(t,\beta) - \epsilon\beta'' \geq 0$. Consider finally the verification that $\epsilon\alpha'' \geq h(t,\alpha)$:

$$\epsilon\alpha'' - h(t,\alpha) = \epsilon u'' + \epsilon w_L'' - h(t,w_L+u)$$

$$- \sum_{j=1}^{2q} \frac{1}{j!} \partial_y^{\,j} h(t,w_L+u) \quad (-\epsilon_1)^j$$

$$- \frac{1}{(2q+1)!} \partial_y^{2q+1} h(t,w_L+u-\theta\epsilon_1)(-\epsilon_1)^{2q+1} \quad .$$

Now by assumptions (1) and (4) there is a solution w_L of (*) with the stated properties and a positive gauge function $\omega = \omega(\epsilon) = O(\sqrt{\epsilon})$ such that on $(a,a+\omega(\epsilon))$

$$\Gamma(t,\epsilon) = \epsilon w_L'' - h(t,w_L+u) - \sum_{j=1}^{2q} \frac{1}{j!} \partial_y^{\,j} h(t,w_L+u)(-\epsilon_1)^j > 0$$

and

$$\Gamma(t,\epsilon) > \epsilon K$$

for any positive constant K. (A function $\omega = \omega(\epsilon)$ which is defined and continuous for all small $\epsilon > 0$ and which satisfies $\lim_{\epsilon\to 0^+} \omega(\epsilon) = 0$ is called a gauge function.) Moreover, on $(a+\omega(\epsilon),b)$ $\Gamma(t,\epsilon) > 0$ and $\Gamma(t,\epsilon) = O(1)$; consequently, for such t $\partial_y^{2q+1} h(t,w_L+u-\theta\epsilon_1) = \partial_y^{2q+1} h(t,u) + \rho(t,\epsilon)$, with $\rho(t,\epsilon) = O(\Gamma)$. For t in $(a,a+\omega(\epsilon))$ we thus have that

$$\epsilon\alpha''-h(t,\alpha) = \epsilon u'' + \Gamma(t,\epsilon) - \partial_y^{2q+1}h[\cdot]\, O(\epsilon)$$

$$\geq \epsilon M + \Gamma(t,\epsilon) - \epsilon M_1 > 0 \quad ;$$

while for t in $(a+\omega(\epsilon),b)$

$$\epsilon\alpha''-h(t,\alpha) = \epsilon u'' + \Gamma(t,\epsilon) - \frac{1}{(2q+1)!}\,\partial_y^{2q+1}h[\cdot]\,(-\epsilon_1)^{2q+1}$$

$$\geq -\epsilon M + \Gamma(t,\epsilon) + \frac{\epsilon\gamma}{(2q+1)!} + \frac{\rho(t,\epsilon)}{(2q+1)!}\,\epsilon\gamma m^{-1} \geq 0$$

for $\gamma \geq (2q+1)!M$ and ϵ sufficiently small since $\rho = O(\Gamma)$. Here $[\cdot] = (t,w_L+u - \theta\epsilon_1)$ for $0 < \theta < 1$. All of the required inequalities are satisfied and we conclude that (P_1) has a solution $y = y(t,\epsilon)$ with the stated properties by Nagumo's theorem, that is, $\alpha(t,\epsilon) \leq y(t,\epsilon) \leq \beta(t,\epsilon)$ for $a \leq t \leq b$. The other cases are handled analogously. We remark that this theorem with $q = 0$, that is $h_y(t,u) > 0$, was first proved by Fife [9] .

We consider next the case when the reduced solution has $(II)_n$-or $(III)_n$-stability on $[a,b]$. These theorems are proved by reasoning as in the proof of Theorem 4.1.

Theorem 4.2. Assume that

(1) the equation $h(t,u) = 0$ has a solution $u = u(t)$ of class $C^{(2)}[a,b]$ with $u(a) \leq A$, $u(b) \leq B$ and $u'' \geq 0$;

(2) the function h is of class $C^{(1)}$ with respect to t and of class $C^{(n)} (n \geq 2)$ with respect to y in $D_1 : a \leq t \leq b$, $y = u(t) + d(t,\epsilon)$, where $d = O(\epsilon^{1/n})$ for $a+\delta \leq t \leq b-\delta$, $d \geq A-u(a)$ for $a \leq t < a+\delta$ and $d \geq B-u(b)$ for $b-\delta < t \leq b$ with δ a small positive constant ;

(3) *the function* u *is* $(II)_n$-*stable on* $[a,b]$;

(4) $\int_{u(a)}^{\eta} h(a,s)ds > 0$ *for* $u(a) < \eta \leq A$ (*if* $u(a) < A$)

and

$$\int_{u(b)}^{\eta} h(b,s)ds > 0 \quad \text{for} \quad u(b) < \eta \leq B \quad (\text{if } u(b) < B) .$$

Then (P_1) *has a solution* $y = y(t,\epsilon)$ *for each sufficiently small* $\epsilon > 0$. *In addition, for* $a \leq t \leq b$

$$u(t) \leq y(t,\epsilon) \leq u(t) + w_L(t,\epsilon) + w_R(t,\epsilon) + O(\epsilon^{1/n})$$

where $w_L(t,\epsilon)(w_R(t,\epsilon))$ *is a positive* $O(\sqrt{\epsilon})$-*boundary layer function at* $t = a$ $(t = b)$.

Theorem 4.3. Assume that

(1) *the equation* $h(t,u) = 0$ *has a solution* $u = u(t)$ *of class* $C^{(2)}[a,b]$ *with* $u(a) \geq A$, $u(b) \geq B$ *and* $u'' \leq 0$;

(2) *the function* h *has the same smoothness as in* Theorem 4.2 ;

(3) *the function* u *is* $(III)_n$-*stable on* $[a,b]$;

(4) $\int_{\eta}^{u(a)} h(a,s)ds < 0$ *for* $A \leq \eta < u(a)$ (*if* $u(a) > A$)

and

$$\int_{\eta}^{u(b)} h(b,s)ds < 0 \quad \text{for} \quad B \leq \eta < u(b) \quad (\text{if } u(b) > B) .$$

Then (P_1) *has a solution* $y = y(t,\epsilon)$ *for each sufficiently small* $\epsilon > 0$. *In addition, for* $a \leq t \leq b$

$$u(t) + w_L(t,\epsilon) + w_R(t,\epsilon) - O(\epsilon^{1/n}) \leq y(t,\epsilon) \leq u(t)$$

where $w_L(t,\epsilon)$ $(w_R(t,\epsilon))$ is a negative $O(\sqrt{\epsilon})$-boundary layer function at $t = a$ $(t = b)$.

In the following theorems we give sufficient conditions for the existence of a solution of (P_1) possessing a shock layer at a point within (a,b) .

Theorem 4.4. Assume that

(1) the equation $h(t,u) = 0$ has two solutions $u = u_L(t)$ and $u = u_R(t)$ which exist and are of class $C^{(2)}$ on $[a,t_L]$ and $[t_R,b]$, respectively, $a \le t_R < t_L \le b$; moreover, $u_L \ne u_R$, $u_L(a) = A$, $u_R(b) = B$;

(2) the function h is of class $C^{(1)}$ with respect to t and of class $C^{(2q+1)}$ $(q \ge 0)$ with respect to y in D_2 : $a \le t \le b$, $y = u_L(t)+d_1(t,\epsilon)$ for $a \le t \le t_L$, $y = u_R(t) + d_2(t,\epsilon)$ for $t_R \le t \le b$, where $d_1 = O(\epsilon^{1/(2q+1)})$ for $a \le t \le t_R$, $d_1 \ge |u_L(t)-u_R(t)|$ for $t_R < t < t_L$ and $d_2 = O(\epsilon^{1/(2q+1)})$ for $t_L \le t \le b$, $d_2 \ge |u_L(t)-u_R(t)|$ for $t_R < t < t_L$;

(3) the functions u_L and u_R are $(I)_q$-stable on their respective intervals of existence ;

(4) there exists a t^* in (t_R,t_L) such that $J[t^*] = 0$ and $(u_R(t^*)-u_L(t^*))J'[t^*] < 0$ where $J[t] = \int_{u_L(t)}^{u_R(t)} h(t,s)\,ds$ if $u_L < u_R$

or

$$J[t] = \int_{u_R(t)}^{u_L(t)} h(t,s)\,ds \quad \text{if} \quad u_L > u_R \;;$$

in addition,

$$\int_{u_L(t^*)}^{\tau} h(t^*,s)\,ds > 0 \quad \text{for} \quad u_L(t^*) < \tau < u_R(t^*) \quad \text{if} \quad u_L < u_R$$

while

$$\int_{u_R(t^*)}^{\tau} h(t^*,s)\,ds > 0 \quad \text{for} \quad u_R(t^*) < \tau < u_L(t^*) \quad \text{if} \quad u_L > u_R \ .$$

Then (P_1) has a solution $y = y(t,\epsilon)$ for each sufficiently small $\epsilon > 0$. In addition,

$$y(t,\epsilon) = u_L(t) + \chi_L(t,\epsilon) + O(\epsilon^{1/(2q+1)}) \quad \text{for} \quad a \le t \le t^* \ ,$$
$$y(t,\epsilon) = u_R(t) + \chi_R(t,\epsilon) + O(\epsilon^{1/(2q+1)}) \quad \text{for} \quad t^* \le t \le b$$

where χ_L, χ_R are $O(\sqrt{\epsilon})$-shock layer functions at $t = t^*$. (That is, $\chi_L(t,\epsilon) = O(|u_L(t^*) - u_R(t^*)|)$ for $t^* - c_1\sqrt{\epsilon} \le t \le t^*$, $\chi_R(t,\epsilon) = O(|u_L(t^*) - u_R(t^*)|)$ for $t^* \le t \le t^* + c_2\sqrt{\epsilon}$ and $\lim_{\epsilon\to 0^+} \chi_{L,R}(t,\epsilon) = 0$ for each fixed t in $[a,b]$, $t \ne t^*$.)

Proof. The idea of the proof is to construct functions α,β like those in the proof of Theorem 4.1 which satisfy the proper inequalities and which bound $y(t,\epsilon)$ in the sense of figure 10.

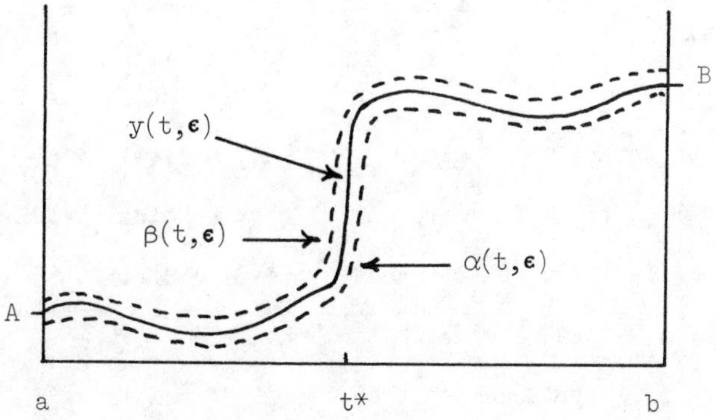

Fig. 10

Bounding Solutions α,β in the case $u_L < u_R$.

Suppose for definiteness that $u_L < u_R$ and define for $\epsilon > 0$ and $\epsilon_1 = (\epsilon\gamma m^{-1})^{1/(2q+1)}$

$$\alpha(t,\epsilon) = \begin{cases} u_L(t)-\epsilon_1 , & a \le t \le t^* , \\ u_R(t)+\chi_R(t,\epsilon) - \epsilon_1 , & t^* \le t \le b , \end{cases}$$

and

$$\beta(t,\epsilon) = \begin{cases} u_L(t) + \chi_L(t,\epsilon) + \epsilon_1 , & a \le t \le t^* , \\ u_R(t) + \epsilon_1 , & t^* \le t \le b . \end{cases}$$

Here $\chi_R(t,\epsilon) < 0$ is a certain solution of the shock layer inequality (*) $\epsilon\tilde{y}'' > h(t^*,\tilde{y} + u_R(t^*))$ satisfying $\chi_R(t^*,\epsilon) = u_L(t^*) - u_R(t^*)$, $\chi_R'(t,\epsilon) > 0$, $\lim_{\epsilon\to 0^+} \chi_R'(t^*,\epsilon) = \infty$ and $\lim_{\epsilon\to 0^+}\chi_R(t,\epsilon) = 0$ for $t > t^*$ while $\chi_L(t,\epsilon) > 0$ is a certain solution of

(**) $\qquad \epsilon\tilde{y}'' < h(t^*,\tilde{y} + u_L(t^*))$

satisfying $\chi_L(t^*,\epsilon) = u_R(t^*) - u_L(t^*)$, $\chi_L'(t,\epsilon) > 0$, $\lim_{\epsilon\to 0^+} \chi_L'(t^*,\epsilon) = \infty$ and $\lim_{\epsilon\to 0^+} \chi_L(t,\epsilon) = 0$ for $t < t^*$. In addition, m is a positive constant such that $\partial_y^{2q+1}h(t,\sigma) > m$ for $\sigma = u_L$ or u_R and γ is a positive constant to be determined. Clearly $\alpha \le \beta$, $\alpha(a,\epsilon) \le A \le \beta(a,\epsilon)$ and $\alpha(b,\epsilon) \le B \le \beta(b,\epsilon)$; however, a slight technical difficulty arises because $\alpha'(t^*-0) < \alpha'(t^*+0)$ and $\beta'(t^*-0) > \beta'(t^*+0)$ for ϵ small. That is, α,β are of class $C^{(2)}$ on $[a,b] - \{t^*\}$. Nagumo's original theory is inapplicable but since α',β' satisfy the above inequalities at $t = t^*$ his technique can be modified to allow bounding solutions with finitely many such "corners" (cf. [16],[18]).

Consider only the verification that α satisfies $\epsilon\alpha'' \geq h(t,\alpha)$ since the verification for β proceeds analogously. On (a,t^*) we have that

$$\epsilon\alpha''-h(t,\alpha) = \epsilon u_L'' - \sum_{j=0}^{2q} \frac{1}{j!} \partial_y^j h(t,u_L)(-\epsilon_1)^j$$

$$- \frac{1}{(2q+1)!} \partial_y^{2q+1} h(t,u_L - \theta\epsilon_1)(-\epsilon_1)^{2q+1}$$

$$\geq -\epsilon M + \frac{\epsilon\gamma}{(2q+1)!} \geq 0$$

for $\gamma \geq (2q+1)!M$. Here $|u_L''(t)| \leq M$ and $0 < \theta < 1$. On (t^*,b) we have that

$$\epsilon\alpha''-h(t,\alpha) = \epsilon u_R'' + \epsilon\chi_R'' - h(t,\chi_R+u_R)$$

$$- \sum_{j=1}^{2q} \frac{1}{j!} \partial_y^j h(t,\chi_R+u_R) (-\epsilon_1)^j$$

$$- \frac{1}{(2q+1)!} \partial_y^{2q+1} h(t,\chi_R + u_R - \theta\epsilon_1)(-\epsilon_1)^{2q+1} .$$

Now by assumptions (1) and (4) there exists a solution $\chi_R = \chi_R(t,\epsilon)$ of (*) with the stated properties and a positive gauge function $\omega = \omega(\epsilon) = O(\sqrt{\epsilon})$ such that on $(t^*,t^* + \omega(\epsilon))$

$$\Gamma(t,\epsilon) = \epsilon\chi_R'' - h(t,\chi_R+u_R) - \sum_{j=1}^{2q} \frac{1}{j!} \partial_y^j h(t,\chi_R+u_R)(-\epsilon_1)^j > 0$$

and $\Gamma(t,\epsilon) > \epsilon K$ for any positive constant K . Moreover, $\Gamma(t,\epsilon) > 0$ and $\Gamma(t,\epsilon) = O(1)$, for t in $(t^* + \omega(\epsilon),b)$; consequently, for such t

$$\partial_y^{2q+1} h(t,\chi_R+u_R - \theta\epsilon_1) = \partial_y^{2q+1} h(t,u_R) + \rho(t,\epsilon)$$

with $\rho(t,\epsilon) = O(\Gamma)$. For t in $(t^*,t^* + \omega(\epsilon))$ it follows that

$$\epsilon\alpha''-h(t,\alpha) = \epsilon u_R'' + \Gamma(t,\epsilon) - \partial_y^{2q+1}h[\cdot]\ O(\epsilon)$$

$$\geq -\epsilon M + \Gamma(t,\epsilon) - \epsilon M > 0$$

while for t in $(t^*+\omega(\epsilon),b)$

$$\epsilon\alpha''-h(t,\alpha) = \epsilon u_R'' + \Gamma(t,\epsilon) - \frac{1}{(2q+1)!}\partial_y^{2q+1}h[\cdot](-\epsilon_1)^{2q+1}$$

$$\geq -\epsilon M + \Gamma(t,\epsilon) + \frac{\epsilon\gamma}{(2q+1)!} + \frac{\rho(t,\epsilon)}{(2q+1)!}\ \epsilon\gamma m^{-1} \geq 0$$

for $\gamma \geq (2q+1)!M$ and ϵ sufficiently small since $\rho(t,\epsilon) = O(\Gamma)$. Thus α satisfies all of the requisite inequalities; similarly β can be shown to do the same. By the modification of Nagumo's theorem noted above we conclude that (P_1) has a solution $y = y(t,\epsilon)$ such that $\alpha(t,\epsilon) \leq y(t,\epsilon) \leq \beta(t,\epsilon)$ for $a \leq t \leq b$.

It also follows that there can be shock layer behavior when, for example, u_L is $(II)_n$-stable and u_R is $(III)_m$-stable. Recall that in such cases we require that $u_L'' \geq 0$ for $a \leq t \leq t^*$ and $u_R'' \leq 0$ for $t^* \leq t \leq b$. More generally we have the next result whose proof can be modelled after that of Theorem 4.4.

Theorem 4.5. Assume that

(1) the reduced equation $h(t,u) = 0$ has two solutions $u = u_L(t)$ and $u = u_R(t)$ of class $C^{(2)}$ on $[a,t_L]$ and $[t_R,b]$, respectively, $a \leq t_R < t_L \leq b$, with $u_L \neq u_R$, $u_L(a) = A$ and $u_R(b) = B$;

(2) the function h is of class $C^{(1)}$ with respect to t and of class $C^{(n)}$ $(n \geq 2)$ with respect to y in D_3, where D_3 is the domain D_2 of Theorem 4.4 with $(2q+1)$ replaced by n;

(3) the function u_L is stable on $[a,t_L]$ and u_R is stable on $[t_R,b]$.

Assume finally that assumption (4) of Theorem 4.4 is satisfied.

Then the conclusion of Theorem 4.4 is valid with $(2q+1)$ replaced by n .

We note that Theorem 4.4 was first proved by Fife [11,12] for the case that u_L and u_R are $(I)_0$-stable, that is, $h_y(t,u_L) > 0$ and $h_y(t,u_R) > 0$.

It is now possible to combine our results concerning boundary and shock layer behavior into one theorem. Namely, stable reduced solutions of (P_1) often support boundary layers at each endpoint as well as shock layers within (a,b) . We simply require these solutions to satisfy the correct boundary layer and shock layer inequalities. The interested reader can formulate the precise results based on the above theorems; however, we do give simple examples of such behavior in Section 7.

Remark 4.1. If there is a single unstable reduced solution between u_L and u_R then the inequality

$$(u_R(t^*) - u_L(t^*)) \int_{u_L(t^*)}^{\tau} h(t^*,s)\,ds > 0$$

for τ between $u_L(t^*)$ and $u_R(t^*)$ is automatically satisfied.

Remark 4.2. We note that if $u_L < u_R$ then any shock layer is an S-shock, that is, $J'[t^*] < 0$, while if $u_L > u_R$ then any shock layer is a Z-shock, that is, $J'[t^*] > 0$ (cf. assumption (4) of Theorem 4.4).

Remark 4.3. It is essential that J is a nonconstant function of t , namely that $h_t \not\equiv 0$. If $h_t \equiv 0$ then our theory is inapplicable.

Shock layer behavior can occur in this case also together with related transition phenomena. Some of these are discussed in [3; Chap. 18] and more fully in [31]. Here the functional J can be interpreted as the potential energy of a system whose dynamics are modelled by (P_1).

Remark 4.4. Similar results have been obtained by Boglaev [1] and Vasil'eva [33] who also used J-functionals of the above form.

Remark 4.5. It is possible to combine the reasonining used in the formulation of the boundary layer theory given in Theorems 4.1-4.3 with the corner layer theory discussed by the author in [16] to extend those results. Alternatively the results given in Theorems 4.1-4.3 can be extended to the case where the reduced solution is only piecewise differentiable on $[a,b]$.

We take this opportunity to correct a false statement made in [16] for the example

$$\epsilon y'' = y^2 - t^2 \;, \quad -1 < t < 1 \;,$$
$$y(-1,\epsilon) = A \;, \quad y(1,\epsilon) = B \;.$$

We stated that this problem had no solution of bounded t-variation if $A < -1$ or $B < -1$. However this is incorrect. Suppose for example that A and B belong to $(-2,1)$, then this problem has a solution $y = y(t,\epsilon)$ of bounded t-variation; indeed,

$$\lim_{\epsilon\to 0^+} y(t,\epsilon) = |t| \quad \text{for} \quad -1 < t < 1$$

because

$$\int_{\eta}^{1}(s^2-1)\,ds < 0 \quad \text{for} \quad -2 < A \le \eta < 1$$

and

$$\int_{\eta}^{1}(s^2-1)\,ds < 0 \quad \text{for} \quad -2 < B \le \eta < 1 \ .$$

If A or B is less than -2 then our original statement in [16] is valid.

We close this section by discussing the connection between our results on boundary and shock layer behavior. Suppose for example that the reduced equation has two solutions u_L and u_R such that $u_L(a) = A$, $u_L(b) < B$, $u_R(b) = B$ and $u_R(a) > A$. Suppose also that there is a single unstable reduced solution between u_L and u_R . By the above theory we know that u_R supports a boundary layer at $t = a$ provided $\int_{\eta}^{u_R(a)} h(a,s)\,ds < 0$ for $A \le \eta < u_R(a)$, which in this case is equivalent to $\Gamma_1 = \int_{A}^{u_R(a)} h(a,s)\,ds < 0$. Similarly u_L supports a boundary layer at $t = b$ provided $\int_{u_L(b)}^{\eta} h(b,s)\,ds > 0$ for $u_L(b) < \eta \le B$, that is, if $\Gamma_2 = \int_{u_L(b)}^{B} h(b,s)\,ds > 0$. Suppose now that <u>neither</u> of these inequalities is satisfied, that is, $\Gamma_1 \ge 0$ and $\Gamma_2 \le 0$. Consider the functional $J[t] = \int_{u_L(t)}^{u_R(t)} h(t,s)\,ds$ and assume that $J'[t] < 0$ for $a \le t \le b$. Then this implies that Γ_1 is strictly positive and Γ_2 is strictly negative. We claim that $J[t^*] = 0$ at a single point t^* in (a,b) for

$$J[a] = \Gamma_1 > 0 \quad \text{and} \quad J[b] = \Gamma_2 < 0$$

and so the existence of t^* follows by continuity. Conversely, suppose that $J'[t] < 0$ and that $J[t^*] = 0$ for some t^* in (a,b). Then $J[a] = \Gamma_1 > 0$ and $J[b] = \Gamma_2 < 0$ and so boundary layer behavior is impossible. Our results are summarized in the following lemma.

Lemma 4.1. *Suppose that* $h(t,u) = 0$ *has the three solutions with the stated properties and suppose that* $J'[t] < 0$ *for* $a \leq t \leq b$. *Then a solution* $y = y(t,\epsilon)$ *of* (P_1) *has an* S-*shock layer in* (a,b) *if and only if* y *does not have boundary layers at either endpoint.*

A similar lemma can be stated for the case of a Z-shock.

Finally we give a sufficient condition for the occurrence of shock layer behavior in the presence of boundary layer behavior. Suppose that the reduced equation has two stable solutions u_1, u_2 and an unstable one $\tilde{u}$ with $u_1 < \tilde{u} < u_2$, $u_1(a) < A < \tilde{u}(a)$ and $u_2(b) > B > \tilde{u}(b)$. Then u_1 supports a boundary layer at $t = a$ if

$$\int_{u_1(a)}^{\eta} h(a,s)\,ds < 0 \quad \text{for} \quad u_1(a) < \eta \leq A$$

while u_2 supports a boundary layer at $t = b$ if

$$\int_{\eta}^{u_2(b)} h(b,s)\,ds < 0 \quad \text{for} \quad B \leq \eta < u_2(b) \quad .$$

Suppose however that u_1 cannot support a boundary layer at $t = b$, that is,

$$\int_{u_1(b)}^{B} h(b,s)\,ds \leq 0$$

and that u_2 cannot support a boundary layer at $t = a$, that is,

$$\int_{A}^{u_2(a)} h(a,s)\,ds \geq 0 \quad .$$

Consider the functional

$$\tilde{J}[t] = \int_{u_1(t)}^{u_2(t)} h(t,s)\,ds$$

and suppose that $\tilde{J}'[t] < 0$ for $a \leq t \leq b$. It follows that $\tilde{J}[t^*] = 0$ at a point t^* in (a,b) since

$$\tilde{J}[a] = \int_{u_1(a)}^{u_2(a)} h(a,s) = \left(\int_{u_1(a)}^{A} + \int_{A}^{u_2(a)}\right) h(a,s)\,ds > 0$$

and

$$\tilde{J}[b] = \int_{u_1(b)}^{u_2(b)} h(b,s)\,ds = \left(\int_{u_1(b)}^{B} + \int_{B}^{u_2(b)}\right) h(b,s)\,ds < 0 \ .$$

Similar remarks apply to the case of a Z-shock.

5. The Problem (P_2). Consider now

$$(P_2) \qquad \begin{aligned} &\epsilon y'' = f(t,y)y' + g(t,y), \quad a < t < b \ , \\ &y(a,\epsilon) = A \ , \quad y(b,\epsilon) = B \ , \end{aligned}$$

and let us first examine solutions which possess boundary layers at $t = a$. (The case of a boundary layer at $t = b$ is handled by making the change of variable $t' = b+a-t$ and applying our results to the transformed problem.) A very general sufficient condition for boundary layer behavior was given many years ago by Coddington and Levinson [5]; however, since their result has been largely overlooked by later writers we give it here.

Theorem 5.1. (Coddington and Levinson) *Assume that*

(1) *the reduced equation* $f(t,u)u' + g(t,u) = 0$ *has a solution* $u = u_R(t)$ *of class* $C^{(2)}[a,b]$ *such that* $u_R(b) = B$;

(2) the functions f and g are of class $C^{(1)}$ with respect to t and y in $\mathcal{R}$: $a \leq t \leq b$, $y = u_R(t) + d(t,\epsilon)$ where $d \geq |A-u_R(a)|$ for $a \leq t < a+\delta$ and $d = O(\epsilon)$ for $a+\delta \leq t \leq b$ with δ a small positive constant;

(3) the function u_R is globally stable, that is, $f(t,u_R(t)) < 0$ for $a \leq t \leq b$;

(4) $\int_{\eta}^{u_R(a)} f(a,s)ds < 0$ for $A < \eta < u_R(a)$ (if $u_R(a) > A$)

or

$$\int_{u_R(a)}^{\eta} f(a,s)ds < 0 \quad \text{for} \quad u_R(a) < \eta \leq A \quad (\text{if } u_R(a) < A) .$$

Then (P_2) has a solution $y = y(t,\epsilon)$ in $\mathcal{R}$ for each sufficiently small $\epsilon > 0$ such that for $a < t \leq b$

$$\lim_{\epsilon\to 0^+} y(t,\epsilon) = u_R(t)$$

and

$$\lim_{\epsilon\to 0^+} y'(t,\epsilon) = u_R'(t) .$$

Under these assumptions Coddington and Levinson showed that this solution is unique in the sense that there is no other solution of (P_2) which satisfies the stated limiting relations.

We note that if $f(a,\eta) < 0$ for all η between $u_R(a)$ and A then this result may be strengthened as follows.

Theorem 5.2. (Wasow [35]; Erdélyi [8]; O'Malley [28]; Chang [4]) Make the same assumptions as in Theorem 5.1 with the exceptions that in (2) f and g are of class $C^{(2)}$ and in (3) $f(t,y) \leq -k < 0$

for all (t,y) *in* $\mathcal{R}$. *Then the conclusion of* Theorem 5.1 *is valid; moreover, for* $a \le t \le b$

$$y(t,\epsilon) = u_R(t) + O(|A-u_R(a)| \exp[-kt\epsilon^{-1}]) + O(\epsilon)$$

and

$$y'(t,\epsilon) = u_R'(t) + O(\epsilon^{-1} \exp[-kt\epsilon^{-1}]) + O(\epsilon) \quad .$$

Finally it turns out that similar order estimates can be given under Coddington and Levinson's original assumptions, namely

Theorem 5.3. Under the assumptions of Theorem 5.1 *we can further conclude that for* $a \le t \le b$

$$y(t,\epsilon) = u_R(t) + w_L(t,\epsilon) + O(\epsilon)$$

and

$$y'(t,\epsilon) = u_R'(t) + w_L'(t,\epsilon) + O(\epsilon)$$

where $w_L(t,\epsilon)$ *is a* $O(\epsilon)$*-boundary layer function at* $t = a$. (*That is,* $w_L(t,\epsilon) = O(|A - u_R(a)|)$ *for* $a \le t \le a+c_1\epsilon$ *and* $\lim_{\epsilon \to 0+} w_L(t,\epsilon) = 0$ *for each fixed* t *in* $(a,b]$.)

Proof. The proof follows that of Theorem 4.1. Suppose for definiteness that $u_R(a) > A$ then define for $\epsilon > 0$ and $a \le t \le b$

$$\alpha(t,\epsilon) = u_R(t) + w_L(t,\epsilon) - \epsilon\gamma\ell^{-1}(\exp[\lambda(t-b)] - 1)$$

and

$$\beta(t,\epsilon) = u_R(t) + \epsilon\gamma\ell^{-1}([\exp \lambda(t-b)] - 1) \quad .$$

Here $w_L(t,\epsilon) < 0$ is a certain solution of the boundary layer inequality (*) $\epsilon\tilde{y}'' > f(a,\tilde{y} + u_R(a))\tilde{y}'$ satisfying $w_L(a,\epsilon) = A-u_R(a)$

and $\lim_{\epsilon\to 0^+} w_L(t,\epsilon) = 0$ for each fixed $t > a$. The positive constant ℓ is an upper bound on

$$f_y(t,y)u_R' + g_y(t,y) \quad \text{for} \quad (t,y) \quad \text{in} \quad \mathcal{R}$$

and

$$\lambda = -\ell k^{-1} + O(\epsilon) < 0$$

is a root of $\epsilon\lambda^2 + k\lambda + \ell$, where $f(t,u_R(t)) \le -k < 0$ for $a \le t \le b$. (The existence of such a k is guaranteed by assumption (3).) Finally γ is a positive constant to be chosen below. The conditions of Nagumo's theorem ([27], [25]) will be satisfied if $\alpha \le \beta$, $\alpha(a,\epsilon) \le A \le \beta(a,\epsilon)$, $\alpha(b,\epsilon) \le B \le \beta(b,\epsilon)$ and for $a < t < b$

$$\epsilon\alpha'' \ge f(t,\alpha)\alpha' + g(t,\alpha) \quad \text{and} \quad \epsilon\beta'' \le f(t,\beta)\beta' + g(t,\beta) \quad .$$

The first three inequalities are certainly satisfied. Consider now β and set $\epsilon_1 = \epsilon_1(t,\epsilon) = \epsilon\gamma\ell^{-1}\exp[\lambda(t-b)]$, then

$$\begin{aligned} f(t,\beta)\beta' + g(t,\beta) - \epsilon\beta'' &= f(t,u_R)u_R' + f(t,u_R)\lambda\epsilon_1 \\ &\quad + f_y[\cdot]\,u_R'(\epsilon_1 - \epsilon\gamma\ell^{-1}) + O(\epsilon^2\gamma) \\ &\quad + g(t,u_R) + g_y[\cdot](\epsilon_1 - \epsilon\gamma\ell^{-1}) - \epsilon u_R'' - \epsilon\lambda^2\epsilon_1 \\ &\ge -k\lambda\epsilon_1 - \ell\epsilon_1 + \epsilon\gamma - \epsilon^2\gamma M_1 - \epsilon M \\ &= \epsilon\gamma - \epsilon M - \epsilon^2\gamma M_1 \ge 0 \end{aligned}$$

if $\gamma = M+1$ and ϵ is sufficiently small. Here $|u_R''(t)| \le M$, $|O(\epsilon^2\gamma)| \le \epsilon^2\gamma M_1$ and $[\cdot] = (t, u_R + \theta(\epsilon_1 - \epsilon\gamma\ell^{-1}))$ for $0 < \theta < 1$. Finally consider α and let $[\cdot\cdot] = (t, u_R + w_L - \theta(\epsilon_1 - \epsilon\gamma\ell^{-1}))$ for $0 < \theta < 1$, then

$$\begin{aligned}\epsilon\alpha'' - f(t,\alpha)\alpha' - g(t,\alpha) = {} & \epsilon u_R'' + \epsilon w_L'' - \epsilon\lambda^2\epsilon_1 \\ & - f(t,w_L+u_R)(w_L'+u_R') \\ & + f(t,w_L+u_R)\lambda\epsilon_1 \\ & - f_y[\cdot\cdot]u_R'(\epsilon_1-\epsilon\gamma\ell^{-1}) \\ & - f_y[\cdot\cdot]w_L'(\epsilon_1-\epsilon\gamma\ell^{-1}) - O(\epsilon^2\gamma) \\ & - g(t,w_L+u_R) - g_y[\cdot\cdot](\epsilon_1-\epsilon\gamma\ell^{-1}) \quad .\end{aligned}$$

By assumptions (1) and (4) there exists a solution $w_L = w_L(t,\epsilon)$ of (*) with the stated properties and a positive gauge function $\omega = \omega(\epsilon) = O(\epsilon)$ such that on $(a, a+\omega(\epsilon))$

$$\Gamma(t,\epsilon) = \epsilon w_L'' - f(t,w_L+u_R)(w_L'+u_R') - f_y[\cdot\cdot]w_L'(\epsilon_1-\epsilon\gamma\ell^{-1}) - g(t,w_L+u_R) > 0$$

and $\Gamma(t,\epsilon) > \epsilon K$ for any positive constant K . In addition on $(a+\omega(\epsilon),b)$ $\Gamma(t,\epsilon) > 0$ and $\Gamma(t,\epsilon) = O(1)$; consequently, for such t $f(t,w_L+u_R) = f(t,u_R) + \rho(t,\epsilon)$ with $\rho(t,\epsilon) = O(\Gamma)$. Thus for t in $(a, a+\omega(\epsilon))$

$$\epsilon\alpha'' - f(t,\alpha)\alpha' - g(t,\alpha) = \epsilon u_R'' + \Gamma(t,\epsilon) + O(\epsilon) > 0$$

while for t in $(a+\omega(\epsilon),b)$

$$\begin{aligned}\varepsilon\alpha'' - f(t,\alpha)\alpha' - g(t,\alpha) &= \varepsilon u_R'' + \Gamma(t,\varepsilon) - \varepsilon\lambda^2\varepsilon_1 \\ &\quad - f(t,w_L+u_R)\lambda\varepsilon_1 \\ &\quad - f_y[\cdot\cdot]u_R'(\varepsilon_1-\varepsilon\gamma\ell^{-1}) - O(\varepsilon^2\gamma) \\ &\quad - g_y[\cdot\cdot](\varepsilon_1-\varepsilon\gamma\ell^{-1}) \\ &\geq -\varepsilon M + \Gamma(t,\varepsilon) - \varepsilon\lambda^2\varepsilon_1 - k\lambda\varepsilon_1 \\ &\quad - |\rho(t,\varepsilon)\lambda\varepsilon_1| - \ell\varepsilon_1 + \varepsilon\gamma - \varepsilon^2\gamma M_1 \\ &\geq 0\end{aligned}$$

since $\gamma = M+1$ and $\varepsilon\lambda^2 + k\lambda + \ell = 0$. Thus by Nagumo's theorem there exists a solution $y = y(t,\varepsilon)$ of (P_2) with $\alpha(t,\varepsilon) \leq y(t,\varepsilon) \leq \beta(t,\varepsilon)$ for $a \leq t \leq b$. The estimate on $y' - u_R'$ can be obtained without difficulty and analogous reasoning can be applied to the case $u_R(a) < A$.

For the convenience of the reader we state explicitly the analog of Theorem 5.3 in the case that there is a boundary layer at $t = b$.

Theorem 5.4. Assume that

(1) the reduced equation has a solution $u = u_L(t)$ of class $C^{(2)}[a,b]$ such that $u_L(a) = A$;

(2) the functions f and g are of class $C^{(1)}$ with respect to t and y in $\mathcal{R}_1$: $a \leq t \leq b$, $y = u_L(t) + d_1(t,\varepsilon)$ where $d_1 \geq |B - u_L(b)|$ for $b-\delta < t \leq b$ and $d_1 = O(\varepsilon)$ for $a \leq t \leq b-\delta$ with δ a small positive constant;

(3) the function u_L is globally stable, that is, $f(t,u_L(t)) > 0$ for $a \leq t \leq b$;

(4) $\int_{\eta}^{u_L(b)} f(b,s)ds > 0$ for $B \leq \eta < u_L(b)$ (if $u_L(b) > B$) or

$\int_{u_L(b)}^{\eta} f(b,s)ds > 0$ for $u_L(b) < \eta \leq B$ (if $u_L(b) < B$) .

Then (P_2) has a solution $y = y(t,\epsilon)$ for each sufficiently small $\epsilon > 0$. In addition, for $a \leq t \leq b$

$$y(t,\epsilon) = u_L(t) + w_R(t,\epsilon) + O(\epsilon)$$

and

$$y'(t,\epsilon) = u_L'(t) + w_R'(t,\epsilon) + O(\epsilon)$$

where $w_R(t,\epsilon)$ is a $O(\epsilon)$-boundary layer function at $t = b$. (That is, $w_R(t,\epsilon) = O(|B - u_L(b)|)$ for $b-c_1\epsilon \leq t \leq b$ and $\lim_{\epsilon\to 0+} w_R(t,\epsilon) = 0$ for each fixed t in $[a,b)$.)

We give next a sufficient condition for the occurrence of shock layer behavior. For simplicity we assume in the next theorem that $f(t,0) = g(t,0) \equiv 0$, namely that $u_s \equiv 0$ is a singular reduced solution.

Theorem 5.5. Assume that

(1) the reduced equation has two solutions $u = u_L(t)$ and $u = u_R(t)$ of class $C^{(2)}$ on $[a,t_L]$, $[t_R,b]$, respectively, $a \leq t_R < t_L \leq b$, with $u_L < 0 < u_R$ or $u_L > 0 > u_R$, $u_L(a) = A$ and $u_R(b) = B$;

(2) the functions f and g are of class $C^{(1)}$ with respect to t and y in $\mathcal{D}_2$: $a \leq t \leq b$, $y = u_L(t) + d_1(t,\epsilon)$ for $a \leq t \leq t_L$ and $y = u_R(t) + d_2(t,\epsilon)$ for $t_R \leq t \leq b$ where $d_1 = O(\epsilon)$ for $a \leq t \leq t_R$, $d_1 \geq |u_L(t) - u_R(t)|$ for $t_R < t < t_L$, $d_2 = O(\epsilon)$ for $t_L \leq t \leq b$ and $d_2 \geq |u_L(t) - u_R(t)|$ for $t_R < t < t_L$;

(3) *the function* u_L *is stable on* $[a,t_L]$, *that is,* $f(t,u_L(t)) > 0$, *and* u_R *is stable on* $[t_R,b]$, *that is,* $f(t,u_R(t)) < 0$;

(4) *there is a point* t^* *in* (t_R,t_L) *such that* $J[t^*] = 0$ *and* $J'[t^*] \neq 0$ *where*

$$J[t] = \int_{u_L(t)}^{u_R(t)} f(t,s)\,ds \quad \text{for} \quad t_R \le t \le t_L \quad (\text{if} \quad u_L < u_R)$$

or

$$J[t] = \int_{u_R(t)}^{u_L(t)} f(t,s)\,ds \quad \text{for} \quad t_R \le t \le t_L \quad (\text{if} \quad u_L > u_R) .$$

Then (P_2) *has a solution* $y = y(t,\epsilon)$ *for each sufficiently small* $\epsilon > 0$. *In addition,*

$$y(t,\epsilon) = u_L(t) + \chi_L(t,\epsilon) + O(\epsilon) \quad \text{for} \quad a \le t \le t^*$$

and

$$y(t,\epsilon) = u_R(t) + \chi_R(t,\epsilon) + O(\epsilon) \quad \text{for} \quad t^* \le t \le b$$

where $\chi_L(t,\epsilon)$ *and* $\chi_R(t,\epsilon)$ *are* $O(\epsilon)$*-shock layer functions at* $t = t^*$. *(That is,* $\chi_L(t,\epsilon) = O(|u_L(t^*) - u_R(t^*)|)$ *for* $t^*-c_1\epsilon \le t \le t^*$, $\chi_R(t,\epsilon) = O(|u_L(t^*) - u_R(t^*)|)$ *for* $t^* \le t \le t^*+c_2\epsilon$ *and* $\lim_{\epsilon\to 0+} \chi_{L,R}(t,\epsilon) = 0$ *for each fixed* t *in* $[a,b]$, $t \neq t^*$ *.)* We remark that the theorem does not exclude the possibility that there are several "shock values" t^* and several corresponding solutions of (P_2) .

This theorem is proved by adapting the proof of Theorem 5.3 in exactly the same way that Theorem 4.1 was used to prove Theorem 4.4. The details are omitted.

It is now of interest to examine the relationship between the non-existence of boundary layer behavior and the existence of shock layer behavior. Suppose first that the stable reduced solutions u_L, u_R exist on $[a,b]$ and satisfy $u_L < 0 < u_R$. Assume also that $J'[t] < 0$ for $a \leq t \leq b$.

If u_R cannot support a boundary layer at $t = a$ then it follows that

$$\int_A^{u_R(a)} f(a,s)\,ds > 0$$

while if u_L cannot support a boundary layer at $t = b$ then

$$\int_{u_L(b)}^{B} f(b,s)\,ds < 0 .$$

Clearly $J[t^*] = 0$ for a t^* in (a,b) because

$$J[a] = \int_{u_L(a)=A}^{u_R(a)} f(a,s)\,ds > 0$$

and

$$J[b] = \int_{u_L(b)}^{u_R(b)=B} f(b,s)\,ds < 0 ,$$

that is, there is an S-shock layer at $t = t^*$. Conversely, if $J[t^*] = 0$ for a t^* in (a,b) and $J'[t] < 0$ for $a \leq t \leq b$ then $J[a] > 0$ and $J[b] < 0$ which preclude the occurrence of boundary layer behavior at $t = a$ and $t = b$. Similar remarks apply if $u_L > 0 > u_R$, that is, the nonoccurrence of boundary layer behavior is equivalent to the occurrence of a Z-shock layer.

Suppose however that $u_L < 0 < u_R$ but $J'[t] > 0$ for

$$J[t] = \int_{u_L(t)}^{u_R(t)} f(t,s)\,ds .$$

Then any shock must be an S-shock. Assume now that u_L supports a boundary layer at $t = b$, that is,

$$\int_{u_L(b)}^{B} f(b,s)\,ds > 0$$

and that u_R supports a boundary layer at $t = a$, that is,

$$\int_{A}^{u_R(a)} f(a,s)\,ds < 0 .$$

These inequalities easily imply that $J[t^*] = 0$ for a t^* in (a,b) since $J[a] < 0$ and $J[b] > 0$. Conversely, suppose that $J[t^*] = 0$ for a t^* in (a,b) and that $J'[t] > 0$ for $a \le t \le b$. Then

$$J[a] = \int_{A}^{u_R(a)} f(a,s)\,ds < 0 \quad \text{and} \quad J[b] = \int_{u_L(b)}^{B} f(b,s)\,ds > 0$$

which imply that

$$\int_{\eta}^{u_R(a)} f(a,s)\,ds < 0 \quad \text{for } \underline{\text{all}} \ \eta \ \text{in} \ [A, u_R(a)) ,$$

and

$$\int_{u_L(b)}^{\eta} f(b,s)\,ds > 0 \quad \text{for } \underline{\text{all}} \ \eta \ \text{in} \ (u_L(b), B] .$$

Thus (under certain assumptions) for a given pair of boundary conditions A and B there can exist several solutions of (P_2) which feature boundary and shock layer behavior.

Suppose now that u_L and u_R $(u_L < u_R)$ are not globally stable but that their domains of stability overlap, that is, $a < t_R < t_L < b$ (cf. Section 2). Under the assumption that $u_s \equiv 0$ is the only singular solution of $f(t,u)u' + g(t,u) = 0$, that is, $f(t,0) = g(t,0) \equiv 0$, it follows that u_L becomes unstable when it crosses u_s and so $u_L(t_L) = 0$, while u_R becomes unstable at t_R, that is, $u_R(t_R) = 0$. Consider now

$$J[t] = \int_{u_L(t)}^{u_R(t)} f(t,s)\,ds \quad \text{for} \quad t \quad \text{in} \quad [t_R, t_L]$$

and assume that $J'[t] < 0$. We claim that $J[t^*] = 0$ for a t^* in (t_R, t_L) because

$$J[t_R] = \int_{u_L(t_R)}^{u_R(t_R)=0} f(t_R,s)\,ds > 0$$

(since $f(t_R,s) > 0$ for $u_L(t_R) \le s < 0$) and

$$J[t_L] = \int_{u_L(t_L)=0}^{u_R(t_L)} f(t_L,s)\,ds < 0$$

(since $f(t_L,s) < 0$ for $0 < s \le u_R(t_L)$) .

Similar reasoning applies when u_L is globally stable and u_R is only stable on $[t_R,b]$ $(t_R > a)$ or when u_R is globally stable and u_L is only stable on $[a,t_L]$ $(t_L < b)$. In such cases there is always an S-shock layer at a point in (t_R,b) or in (a,t_L) .

Suppose next that $t_L < t_R$. (If $t_L = t_R$ then $u_L(t_L) = u_R(t_R) = 0$ and so $u_L \equiv u_R$.) The functions u_L and u_R cannot of course support a shock layer in (a,b) ; however, let us assume that $u_s \equiv 0$ is a stable (singular) solution of the reduced equation. It is reasonable to expect that a solution of (P_2) exists and follows the "broken-line" path

$$u(t) = \begin{cases} u_L(t) \ , & a \le t \le t_L \ , \\ 0 \ , & t_L \le t \le t_R \ , \\ u_R(t) \ , & t_R \le t \le b \ , \end{cases}$$

since each subarc is stable. This type of interior crossing phenomenon (involving two stable reduced solutions) was first studied by Haber and Levinson [14] and we will have more to say about their result in the next section. More recent work is contained in [20] whose techniques can be used to prove theorems like the following.

Theorem 5.6. Assume that

(1) the reduced equation has three solutions $u = u_L(t)$, $u = u_s(t)$ and $u = u_R(t)$ defined and of class $C^{(2)}$ on $[a,t_L]$,$[t_L,t_R]$, $[t_R,b]$, respectively, $a < t_L < t_R < b$, with $u_L(a) = A$, $u_L(t_L) = u_s(t_L)$, $u_L'(t_L) \neq u_s'(t_L)$, $u_s(t_R) = u_R(t_R)$, $u_s'(t_R) \neq u_R'(t_R)$ and $u_R(b) = B$;

(2) the functions f and g are of class $C^{(1)}$ with respect to t and of class $C^{(2q+1)}(q \geq 0)$ with respect to y in $\mathcal{R}_2$: $a \leq t \leq b$, $y = u_L(t) + O(\epsilon^{1/(2q+1)})$ for $a \leq t \leq t_L$, $y = u_s(t) + O(\epsilon^{1/2q+1)})$ for $t_L \leq t \leq t_R$ and $y = u_R(t) + O(\epsilon^{1/(2q+1)})$ for $t_R \leq t \leq b$;

(3) for t in $[a,t_L)$ $f(t,u_L(t)) > 0$ and $f(t_L,u_L(t_L)) = 0$; for t in $[t_L,t_R]$ u_s is $(I)_q$-stable ; for t in $(t_R,b]$ $f(t,u_R(t)) < 0$ and $f(t_R,u_R(t_R)) = 0$.

Then (P_2) has a solution $y = y(t,\epsilon)$ for each sufficiently small $\epsilon > 0$. In addition, for $u = u(t)$ as above and $a \leq t \leq b$

$$y(t,\epsilon) = u(t) + O(\epsilon^{1/(2q+1)}) \quad .$$

The question naturally arises as to what can occur in the above situation when the singular intermediate solution ($u_s \equiv 0$ in our case) is unstable. General results do not appear possible; however, for certain problems a solution $y = y(t,\epsilon)$ of (P_2) exists and satisfies

$$\lim_{\epsilon \to 0^+} y(t,\epsilon) = \tilde{u}(t) \quad \text{for} \quad a < t < b$$

where $\tilde{u}(t)$ is a reduced solution which satisfies neither boundary condition but which is stable with respect to boundary layer behavior at each endpoint. (Hereafter when such a situation obtains we shall say that $\tilde{u}$ supports a boundary layer at each endpoint.) An example of this is given in Section 8. We note that such twin layer behavior can also occur when two (globally) unstable solutions are separated by an unstable singular one; an illustration is presented in Section 8.

We conclude this section by noting that stable singular reduced solutions of (P_2) can support boundary layers at each endpoint provided the correct convexity conditions are satisfied there (cf. Section 3). Proceeding as in [18] we can prove the following result.

Theorem 5.7. *Assume that*

(1) *the reduced equation has a singular solution* $u = u_s(t)$ (*that is,* $f(t,u_s(t)) = g(t,u_s(t)) \equiv 0$) *of class* $C^{(2)}[a,b]$ *such that* $f(a,A)\cdot(A - u_s(a)) \leq 0$ *and* $f(b,B)(B - u_s(b)) \geq 0$;

(2) *the functions* f *and* g *are continuous in* t,y *and of class* $C^{(2q+1)}$ $(q \geq 0)$ *with respect to* y *in* $\mathfrak{D}$: $a \leq t \leq b$, $y = u_s(t) + d(t,\epsilon)$ *where* $d \geq |A - u_s(a)|$ *for* $a \leq t < a+\delta$, $d = O(\epsilon^{1/(2q+1)})$ *for* $a+\delta \leq t \leq b-\delta$ *and* $d \geq |B - u_s(b)|$ *for* $b-\delta < t \leq b$ *with* δ *a small positive constant*;

(3) *the function* u_s *is* $(I)_q$-*stable on* $[a,b]$.

Then (P_2) *has a solution* $y = y(t,\epsilon)$ *for each sufficiently small* $\epsilon > 0$. *In addition, for* $a \leq t \leq b$

$$y(t,\epsilon) = u_s(t) + O(|A - u_s(a)|\exp[-(t-a)\epsilon^{-1/2}])$$

$$+ O(|B - u_s(b)|\exp[-(b-t)\epsilon^{-1/2}]) + O(\epsilon) \quad ,$$

<u>if</u> $q = 0$; <u>while if</u> $q \geq 1$

$$y(t,\epsilon) = u_s(t) + O(|A - u_s(a)|(1 + (t-a)\epsilon^{-1/2})^{-1/q})$$

$$+ O(|B - u_s(b)|(1 + (b-t)\epsilon^{-1/2})^{-1/q}) + O(\epsilon^{1/(2q+1)}) \quad .$$

Finally it is possible to combine the boundary layer result just given with the corner layer result given in Theorem 5.6. Namely, it often happens for example that a reduced solution u_R intersects a stable singular reduced solution at t_R in (a,b). If $f(a,A)\cdot(A - u_s(a)) \leq 0$ then (P_2) can be shown to have a solution $y = y(t,\epsilon)$ such that

$$\lim_{\epsilon\to 0^+} y(t,\epsilon) = \begin{cases} u_s(t) , & a < t \leq t_R , \\ u_R(t) , & t_R \leq t \leq b . \end{cases}$$

Similarly if a solution u_L intersects u_s at t_L in (a,b) and if $f(b,B)(B - u_s(b)) \geq 0$ then (P_2) can be shown to have a solution $y = y(t,\epsilon)$ such that

$$\lim_{\epsilon\to 0^+} y(t,\epsilon) = \begin{cases} u_L(t) , & a \leq t \leq t_L , \\ u_s(t) , & t_L \leq t < b . \end{cases}$$

6. The Problem (P_3). We turn now to a discussion of the types of behavior which solutions of

$$(P_3) \quad \begin{aligned} &\epsilon y'' = p(t,y)y'^2 + q(t,y)y' + r(t,y) \quad , \quad a < t < b \quad , \\ &y(a,\epsilon) = A \ , \quad y(b,\epsilon) = B \end{aligned}$$

can exhibit. A detailed study of (P_3) has been given by the author in the case that $p(t,y) \neq 0$ ([18], [22]) and also in the case that $p(t_0,y) = 0$ for all y and a t_0 in (a,b) ([19]). Here we allow $p(t,y)$ to have zeros in y for fixed t which as we have seen in Section 3 is a necessary condition for the occurrence of S- and Z-boundary and shock layer behavior.

Consider first the possibility of boundary layer behavior at $t = a$ and suppose that the reduced equation $p(t,u)u'^2 + q(t,u)u' + r(t,u) = 0$ has a globally stable solution $u = u_R(t)$ such that $u_R(b) = B$ and $u_R(a) > A$. If $p(a,\eta) < 0$ for all η in $[A,u_R(a))$ it is clear that a solution $y = y(t,\epsilon)$ of (P_3) which satisfies

$$\lim_{\epsilon\to 0^+} y(t,\epsilon) = u_R(t) \quad \text{for} \quad a < t \leq b$$

has a $\cap$-boundary layer at $t = a$. This result was proved in [18]; more generally we have

Theorem 6.1. Assume that

(1) the reduced equation has a solution $u = u_R(t)$ of class $C^{(2)}[a,b]$ such that $u_R(b) = B$;

(2) the functions p,q and r are of class $C^{(1)}$ with respect to t and y in $\mathfrak{F}$: $a \le t \le b$, $y = u_R(t) + d(t,\epsilon)$ where $d \ge |A - u_R(a)|$ for $a \le t < a+\delta$ and $d = O(\epsilon)$ for $a+\delta \le t \le b$ with δ a small positive constant ;

(3) the function u_R is globally stable, that is,

$$2p(t,u_R(t))\, u_R'(t) + q(t,u_R(t)) < 0 \quad \text{for} \quad a \le t \le b \ ;$$

(4) $$\int_{\eta}^{u_R(a)} p(a,s)\,ds < 0 \quad \text{for} \quad A \le \eta < u_R(a) \quad (\text{if} \ u_R(a) > A)$$

or

$$\int_{u_R(a)}^{\eta} p(a,s)\,ds > 0 \quad \text{for} \quad u_R(a) < \eta \le A \quad (\text{if} \ u_R(a) < A) \ .$$

Then (P_3) has a solution $y = y(t,\epsilon)$ for each sufficiently small $\epsilon > 0$. In addition, for $a \le t \le b$

$$y(t,\epsilon) = u_R(t) + w_L(t,\epsilon) + O(\epsilon)$$

where $w_L(t,\epsilon)$ is a $O(\epsilon)$-boundary layer function at $t = a$.

Proof. Since $p(t,y)y'^2 + q(t,y)y' + r(t,y) = O(|y'|^2)$ as $|y'| \to \infty$ for (t,y) in $\mathfrak{F}$, Nagumo's theorem ([27], [25]) applies provided we can construct functions α,β such that $\alpha \le \beta$, $\alpha(a,\epsilon) \le A \le \beta(a,\epsilon)$, $\alpha(b,\epsilon) \le B \le \beta(b,\epsilon)$ and for $a < t < b$

$$\epsilon\alpha'' \ge p(t,\alpha)\alpha'^2 + q(t,\alpha)\alpha' + r(t,\alpha)$$

and

$$\epsilon\beta'' \le p(t,\beta)\beta'^2 + q(t,\beta)\beta' + r(t,\beta) \ .$$

Since the proof is only a repetition of that of Theorem 5.3 we content ourselves with simply defining the proper bounding functions in the case that $u_R(a) < A$. Namely, for $\epsilon > 0$ and $a \le t \le b$ define

$$\alpha(t,\epsilon) = u_R(t) - \epsilon\gamma\ell^{-1}(\exp[\lambda(t-b)] - 1)$$

and

$$\beta(t,\epsilon) = u_R(t) + w_L(t,\epsilon) + \epsilon\gamma\ell^{-1}(\exp[\lambda(t-b)] - 1) \quad .$$

Here ℓ is an upper bound on $p_y(t,y)u_R'^2 + q_y(t,y)u_R' + r_y(t,y)$ for (t,y) in $\mathfrak{F}$, γ is a positive constant to be chosen sufficiently large and $\lambda = -\ell k^{-1} + O(\epsilon) < 0$ is a root of $\epsilon\lambda^2 + k\lambda + \ell$, where

$$2p(t,u_R)u_R' + q(t,u_R) < -k < 0 \quad \text{for} \quad a \le t \le b \quad .$$

In addition, $w_L(t,\epsilon) > 0$ is a certain solution of $\epsilon\tilde{y}'' < p(a,\tilde{y} + u_R(a))\tilde{y}'^2$ satisfying $w_L(a,\epsilon) = A - u_R(a)$ and $\lim_{\epsilon\to 0^+} w_L(t,\epsilon) = 0$ for fixed $t > a$. The case $u_R(a) > A$ is handled similarly.

The corresponding result at $t = b$ follows by making the change of variable $t' = a+b-t$ and applying Theorem 6.1 to the transformed problem. For the convenience of the reader we state this explicitly as

Theorem 6.2. Assume that

(1) *the reduced equation has a solution* $u = u_L(t)$ *of class* $C^{(2)}[a,b]$ *such that* $u_L(a) = A$;

(2) *the functions* p,q *and* r *are of class* $C^{(1)}$ *with respect to* t *and* y *in* $\mathfrak{F}_1$: $a \le t \le b$, $y = u_L(t) + d_1(t,\epsilon)$ *where* $d_1 = O(\epsilon)$ *for* $a \le t \le b-\delta$ *and* $d_1 \ge |B - u_L(b)|$ *for* $b-\delta < t \le b$ *with* δ *a small positive constant* ;

(3) *the function* u_L *is globally stable, that is,* $2p(t,u_L(t))u_L'(t) + q(t,u_L(t)) > 0$ *for* $a \le t \le b$;

(4) $\int_{\eta}^{u_L(b)} p(b,s)\,ds < 0$ for $B \leq \eta < u_L(b)$ (if $u_L(b) > B$)

or

$\int_{u_L(b)}^{\eta} p(b,s)\,ds > 0$ for $u_L(b) < \eta \leq B$ (if $u_L(b) < B$) .

Then (P_3) has a solution $y = y(t,\epsilon)$ for each sufficiently small $\epsilon > 0$. In addition,

$$y(t,\epsilon) = u_L(t) + w_R(t,\epsilon) + O(\epsilon) \quad \text{for} \quad a \leq t \leq b$$

where $w_R(t,\epsilon)$ is a $O(\epsilon)$-boundary layer function at $t = b$.

We consider now the existence of solutions of (P_3) which possess shock layer behavior. For simplicity we assume that the reduced solutions u_L and u_R are separated by the singular reduced solution $u_s \equiv 0$.

Theorem 6.3. Assume that

(1) the reduced equation has two solutions $u = u_L(t)$ and $u = u_R(t)$ which are of class $C^{(2)}$ on $[a,t_L]$, $[t_R,b]$, respectively, $a \leq t_R < t_L \leq b$, with $u_L < 0 < u_R$ (or $u_L > 0 > u_R$) and $u_L(a) = A$, $u_R(b) = B$;

(2) the functions p,q and r are of class $C^{(1)}$ with respect to t and y in $\mathfrak{F}_2$: $a \leq t \leq b$, $y = u_L(t) + d_1(t,\epsilon)$ for $a \leq t \leq t_L$ and $y = u_R(t) + d_2(t,\epsilon)$ for $t_R \leq t \leq b$ where $d_1 = O(\epsilon)$ for $a \leq t \leq t_R$, $d_1 \geq |u_L(t) - u_R(t)|$ for $t_R < t < t_L$ and $d_2 \geq |u_L(t) - u_R(t)|$ for $t_R < t < t_L$, $d_2 = O(\epsilon)$ for $t_L \leq t \leq b$;

(3) _the function_ u_L _is stable on_ $[a,t_L]$ _and_ u_R _is stable on_ $[t_R,b]$;

(4) _there is a point_ t^* _in_ (t_R,t_L) _such that_ $J[t^*] = 0$ _and_ $(u_R(t^*) - u_L(t^*))\, J'[t^*] < 0$ _for_

$$J[t] = \int_{u_L(t)}^{u_R(t)} p(t,s)\,ds \quad (\text{if } u_L < u_R)$$

_or_

$$J[t] = \int_{u_R(t)}^{u_L(t)} p(t,s)\,ds \quad (\text{if } u_L > u_R) \ .$$

Then (P_3) _has a solution_ $y = y(t,\epsilon)$ _for each sufficiently small_ $\epsilon > 0$. _In addition,_

$$y(t,\epsilon) = u_L(t) + \chi_L(t,\epsilon) + O(\epsilon) \quad \text{for} \quad a \le t \le t^* \ ,$$

$$y(t,\epsilon) = u_R(t) + \chi_R(t,\epsilon) + O(\epsilon) \quad \text{for} \quad t^* \le t \le b \ ,$$

where $\chi_L(t,\epsilon)$, $\chi_R(t,\epsilon)$ _are_ $O(\epsilon)$-_shock layer functions at_ $t = t^*$.

Proof. The proof is patterned after those of the previous shock layer theorems. Suppose for instance that $u_L < 0 < u_R$ and define for $\epsilon > 0$

$$\alpha(t,\epsilon) = \begin{cases} u_L(t) - \epsilon\gamma\ell^{-1}(\exp[\lambda(a-t)] - 1) \ , & a \le t \le t^* \ , \\ u_R(t) + w_R(t,\epsilon) - \epsilon\gamma\ell^{-1}(\exp[\lambda(t-b)] - 1) \ , & t^* \le t \le b \ , \end{cases}$$

and

$$\beta(t,\epsilon) = \begin{cases} u_L(t) + w_L(t,\epsilon) + \epsilon\gamma\ell^{-1}(\exp[\lambda(a-t)] - 1) \ , & a \le t \le t^* \ , \\ u_R(t) + \epsilon\gamma\ell^{-1}(\exp[\lambda(t-b)] - 1) \ , & t^* \le t \le b \ . \end{cases}$$

Here $\chi_L(t,\epsilon) > 0$ is a certain solution of $\epsilon\tilde{y}'' < p(t^*,\tilde{y} + u_L(t^*))\tilde{y}'^2$ satisfying $\chi_L(t^*,\epsilon) = u_R(t^*) - u_L(t^*)$, $\lim_{\epsilon\to 0^+} \chi_L'(t^*,\epsilon) = \infty$ and $\lim_{\epsilon\to 0^+} \chi_L(t,\epsilon) = 0$ for $t < t^*$, while $\chi_R(t,\epsilon) < 0$ is a certain solution of $\epsilon\tilde{y}'' > p(t^*,\tilde{y} + u_R(t^*))\tilde{y}'^2$ satisfying $\chi_R(t^*,\epsilon) = u_L(t^*) - u_R(t^*)$, $\lim_{\epsilon\to 0^+} \chi_R'(t^*,\epsilon) = \infty$ and $\lim_{\epsilon\to 0^+} \chi_R(t,\epsilon) = 0$ for $t > t^*$.

In the now familiar manner it can be shown that α and β satisfy the correct inequalities. The details are omitted.

It is possible now to show the connection between the nonexistence of boundary layer behavior and the existence of various forms of interior layer phenomena. Suppose for example that $u_L < u_R$ are stable on $[a,b]$ and are separated by the singular reduced solution $u_s \equiv 0$. If u_R cannot support a boundary layer at $t = a$ then

$$\int_A^{u_R(a)} p(a,s)\,ds > 0$$

while if u_L cannot support a boundary layer at $t = b$ then

$$\int_{u_L(b)}^{B} p(b,s)\,ds < 0 \ ;$$

we assume of course that $J'[t] < 0$ for $a \le t \le b$. These inequalities imply that $J[a] > 0$ and $J[b] < 0$ and therefore, $J[t^*] = 0$ for a point t^* in (a,b) . Conversely if $J[t^*] = 0$ for $a < t^* < b$ then $J[a] > 0$ and $J[b] < 0$ and so there cannot be any boundary layer behavior. We remark that the boundary layer inequalities can only be violated when $A < 0$ and $B > 0$. Suppose next that u_L is only stable on $[a,t_L]$ for $t_L < b$ and that u_R is only stable on $[t_R,b]$ for $t_R > a$ with $t_L > t_R$. Since u_L and u_R are separated by the singular solution $u_s \equiv 0$ it follows that $u_L(t_L) = 0$ and $u_R(t_R) = 0$. Arguing as in the case of (P_2) we can show that

$$J[t_R] = \int_{u_L(t_R)}^{u_R(t_R)=0} p(t_R,s)\,ds > 0$$

and

$$J[t_L] = \int_{u_L(t_L)=0}^{u_R(t_L)} p(t_L,s)\,ds < 0$$

and therefore, $J[t^*] = 0$ at a point t^* in (t_R,t_L), that is, there is an S-shock layer at t^*.

Similarly if u_R is globally stable and u_L is only stable on $[a,t_L] \subseteq [a,b]$ or if u_L is globally stable and u_R is only stable on $[t_R,b] \subseteq [a,b]$ then there is an S-shock in (a,t_L) (or in (t_R,b)). Consider for example the first case and look at

$$J[t] = \int_{u_L(t)}^{u_R(t)} p(t,s)\,ds \quad \text{for} \quad a \leq t \leq t_L \ .$$

If boundary layer behavior is not possible at $t = a$ then

$$\int_A^{u_R(a)} p(a,s)\,ds > 0 \quad \text{while} \quad \int_{u_L(t_L)=0}^{u_R(t_L)} p(t_L,s)\,ds < 0$$

since $p(t_L,s) < 0$ for $0 < s \leq u_R(t_L)$. Consequently $J[a] > 0$ and $J[b] < 0$ and so $J[t^*] = 0$ for a t^* in (a,t_L).

Suppose however that $t_L < t_R$ (if $t_L = t_R$ then $u_L \equiv u_R$). Then u_L,u_R cannot support a shock layer in (a,b). If $u_s \equiv 0$ is stable then the broken-line path

$$u(t) = \begin{cases} u_L(t), & a \le t \le t_L , \\ 0 , & t_L \le t \le t_R , \\ u_R(t), & t_R \le t \le b , \end{cases}$$

is stable and it is reasonable to expect that (P_3) has a solution $y = y(t,\epsilon)$ such that $\lim_{\epsilon\to 0^+} y(t,\epsilon) = u(t)$ for $a \le t \le b$. The precise result is contained in the next theorem whose proof can be patterned after those in [20].

Theorem 6.4. *Assume that*

(1) *the reduced equation has three solutions* $u = u_L(t)$, $u = u_s(t)$ *and* $u = u_R(t)$ *defined and of class* $C^{(2)}$ *on* $[a,t_L]$, $[t_L,t_R]$, $[t_R,b]$, *respectively,* $a < t_L < t_R < b$, *with* $u_L(a) = A$, $u_L(t_L) = u_s(t_L)$, $u_L'(t_L) \neq u_s'(t_L)$, $u_s(t_R) = u_R(t_R)$, $u_s'(t_R) \neq u_R'(t_R)$ *and* $u_R(b) = B$;

(2) *the functions* p,q *and* r *are of class* $C^{(1)}$ *with respect to* t *and of class* $C^{(2q+1)}$ $(q \ge 0)$ *with respect to* y *in* $\mathfrak{F}_3 :\ a \le t \le b$, $y = u_L(t) + O(\epsilon^{1/(2q+1)})$ *for* $a \le t \le t_L$, $y = u_s(t) + O(\epsilon^{1/(2q+1)})$ *for* $t_L \le t \le t_R$ *and* $y = u_R(t) + O(\epsilon^{1/(2q+1)})$ *for* $t_R \le t \le b$;

(3) *for* t *in* $[a,t_L)$ $\varphi_L(t) = 2p(t,u_L(t))u_L'(t) + q(t,u_L(t)) > 0$ *and* $\varphi_L(t_L) = 0$; *for* t *in* $[t_L,t_R]$ u_s *is* $(I)_q$-*stable*; $\varphi_R(t_R) = 0$ *and* $\varphi_R(t) < 0$ *for* t *in* $(t_R,b]$.

Then (P_3) *has a solution* $y = y(t,\epsilon)$ *for each sufficiently small* $\epsilon > 0$. *In addition, for* $a \le t \le b$

$$y(t,\epsilon) = u(t) + O(\epsilon^{1/(2q+1)}) .$$

As in the case of (P_2) it can happen that the intermediate singular solution is unstable. Solutions of (P_3) often follow solutions of the reduced equation which satisfy neither boundary condition but which are stable with respect to boundary layer behavior at each endpoint. Similarly if u_L and u_R are (globally) unstable and are separated by an unstable u_s then such twin boundary layer behavior often occurs. An example is given in Section 9.

The final two phenomena we study are related to the stability of a singular reduced solution and to the impossibility of boundary layer behavior. First suppose that the reduced equation has a singular solution u_s which is compatible with the convexity requirements imposed by the boundary conditions. It is not unreasonable to expect that (P_3) has a solution $y = y(t,\epsilon)$ such that $\lim_{\epsilon\to 0^+} y(t,\epsilon) = u_s(t)$ for $a < t < b$. Indeed we have the following theorem which was proved in [18].

Theorem 6.5. *Assume that*

(1) *the reduced equation has a singular solution* $u = u_s(t)$ *of class* $C^{(2)}[a,b]$ *such that* $p(a,A)(A - u_s(a)) \geq 0$ *and* $p(b,B)(B - u_s(b)) \geq 0$;

(2) *the functions* p,q *and* r *are continuous in* t *and of class* $C^{(2q+1)} (q \geq 0)$ *with respect to* y *in* $\mathfrak{F}_4 : a \leq t \leq b$, $y = u_s(t) + d(t,\epsilon)$ *where* $d \geq |A - u_s(a)|$ *for* $a \leq t < a+\delta$, $d = O(\epsilon^{1/(2q+1)})$ *for* $a+\delta \leq t \leq b-\delta$ *and* $d \geq |B - u_s(b)|$ *for* $b-\delta < t \leq b$ *with* δ *a small positive constant* ;

(3) *the function* u_s *is* $(I)_q$-*stable on* $[a,b]$.

Then (P_3) _has a solution_ $y = y(t,\epsilon)$ _for each sufficiently small_ $\epsilon > 0$. _In addition, for_ $a \leq t \leq b$

$$y(t,\epsilon) = u_s(t) + O(|A-u_s(a)|\exp[-(t-a)\,\epsilon^{-1/2}])$$

$$+ O(|B-u_s(b)|\exp[-(b-t)\,\epsilon^{-1/2}]) + O(\epsilon), \text{ if } q = 0 ;$$

while if $q \geq 1$

$$y(t,\epsilon) = u_s(t) + O(|A-u_s(a)|(1 + (t-a)\,\epsilon^{-1/2})^{-1/q})$$

$$+ O(|B-u_s(b)|(1 + (b-t)\,\epsilon^{-1/2})^{-1/q}) + O(\epsilon^{1/(2q+1)}) .$$

Finally suppose for example that $p(t,y) < 0$ for $a \leq t \leq b$ and all y of interest and that the reduced equation has two globally stable solutions u_L and u_R . Since $p(t,y) < 0$ any solution of (P_3) is concave inside of a boundary layer (cf. Section 3), that is, all boundary layers are $\cap$-layers. Suppose however that $u_R(a) < A$ and $u_L(b) < B$ then neither u_L nor u_R can support a $\cap$-layer. It follows that u_L intersects u_R at a point t_0 in (a,b) ; see figure 11.

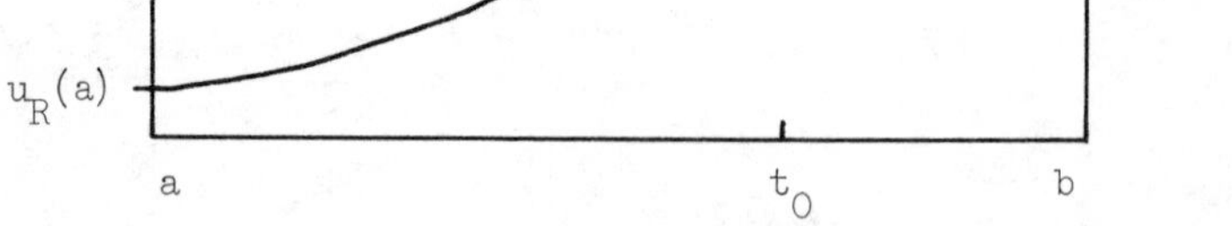

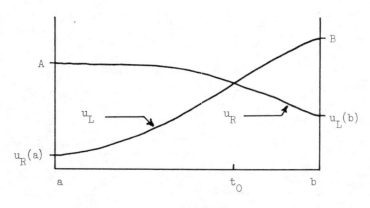

Fig. 11

Interior Crossing of u_L and u_R at $t = t_0$.

In addition, since $p(t,y) < 0$, $u_L'(t_0) < u_R'(t_0)$ and t_0 is the only point of intersection. (We exclude the trivial case $u_L \equiv u_R$.) Under these circumstances (P_3) always has a solution $y = y(t,\epsilon)$ satisfying

$$(\sim) \qquad \lim_{\epsilon\to 0^+} y(t,\epsilon) = \begin{cases} u_L(t) , & a \le t \le t_0 , \\ u_R(t) , & t_0 \le t \le b . \end{cases}$$

To see this we note that the well-known theorem of Haber and Levinson [14] guarantees the existence of such a solution provided

$$\Lambda(\nu) = F(t_0,u_L(t_0) = u_R(t_0),\nu) > 0$$

for all ν in $(u_L'(t_0), u_R'(t_0))$, where $F(t,y,y') = p(t,y)y'^2 + q(t,y)y' + r(t,y)$. However $\Lambda(u_L'(t_0)) = \Lambda(u_R'(t_0)) = 0$ and $\Lambda''(\nu) < 0$ since $p(t,y) < 0$; therefore $\Lambda(\nu) > 0$ for $u_L'(t_0) < \nu < u_R'(t_0)$.

Finally it can happen that $u_L(t_0) = u_R(t_0) = u_s(t_0)$ where u_s is a singular reduced solution, that is, u_L and u_R lose stability precisely when they intersect each other. In terms of the hypotheses of Haber and Levinson's theorem [14] this implies that $F_{y'}(t_0,u_L(t_0),u_L'(t_0)) = F_{y'}(t_0,u_R(t_0),u_R'(t_0)) = 0$ and $F(t_0,u_L(t_0) = u_R(t_0) = u_s(t_0),u_s'(t_0)) = 0$, and so their theorem is inapplicable. If however u_s is stable then it is possible to proceed as in [20] to show that (P_3) has a solution $y = y(t,\epsilon)$ satisfying $(\sim)$. An example of such behavior is given in Example (E5) below.

With this we conclude our discussion of some of the principal types of behavior which solutions of (P_1), (P_2) and (P_3) can exhibit. To motivate this theory and demonstrate the naturalness of our approach we consider in the next sections several examples in detail. Furthermore, to present a balanced account we discuss three examples which are not completely amenable to our techniques.

7. Examples of (P_1). Consider first the problem

$$\text{(E1)}\qquad \begin{aligned} &\epsilon y'' = y(y-1)(y-(t+\tfrac{3}{2})) = h(t,y) , \quad 0 < t < 1 , \\ &y(0,\epsilon) = A , \quad y(1,\epsilon) = B . \end{aligned}$$

The reduced equation $h(t,u) = 0$ has the three solutions $u_1 \equiv 0$, $\tilde{u} \equiv 1$ and $u_2(t) = t + \frac{3}{2}$. Since

$$h_y(t,0) = t + \tfrac{3}{2} > 0 , \; h_y(t,1) = -(t+\tfrac{1}{2}) < 0 \quad \text{and}$$

$$h_y(t,t+\tfrac{3}{2}) = (t+\tfrac{3}{2})(t+\tfrac{1}{2}) > 0 ,$$

we see that u_1 and u_2 are stable and that $\tilde{u}$ is unstable. Consider first the possibility that u_1 supports boundary layers at the endpoints.

Since $h(t,y) < 0$ for $y < 0$ it follows that

$$\int_\eta^{u_1(0)=0} h(0,s)\,ds < 0 \quad \text{for all } \eta < 0$$

and

$$\int_\eta^{u_1(1)=0} h(1,s)\,ds < 0 \quad \text{for all } \eta < 0 .$$

Consequently $u_1 \equiv 0$ supports a concave boundary layer at $t = 0$ $(t = 1)$ if $A < 0$ $(B < 0)$. Suppose next that $A > 0$ and consider

$$\Psi_0(\eta) = \int_{u_1(0)=0}^{\eta} h(0,s)ds \quad \text{for} \quad 0 < \eta \le A \ .$$

A short computation shows that

$$\Psi_0(\eta) = \frac{1}{12}\eta^2(3\eta^2 - 10\eta + 9) > 0 \quad \text{for } \underline{\text{all}} \quad \eta > 0$$

and so $u_1 \equiv 0$ supports three types of boundary layers at $t = 0$. Namely, for $0 < A \le 1$ there is a U-layer, for $1 < A \le \frac{3}{2}$ there is a Z-layer and for $A > \frac{3}{2}$ there is a convex-concave-convex layer, as shown in figure 12.

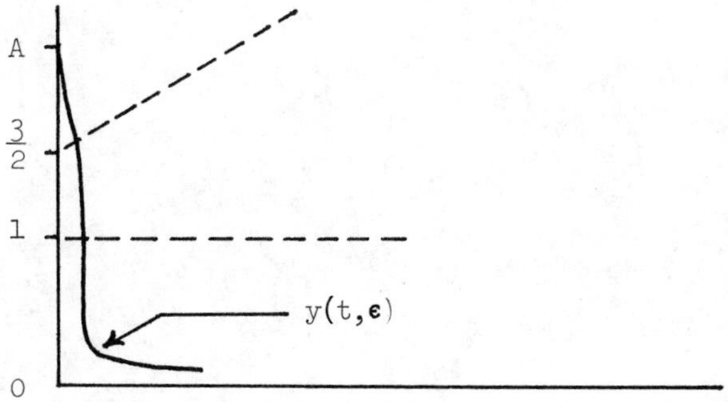

Fig. 12

Convex-Concave-Convex Boundary Layer at $t = a$.

Similarly if $B > 0$ we examine

$$\Psi_1(\eta) = \int_0^{\eta} h(1,s)ds \quad \text{for} \quad 0 < \eta \le B \ .$$

A short computation again shows that

$$\Psi_1(\eta) = \frac{1}{12}\eta^2(3\eta^2 - 14\eta + 15) = \frac{1}{12}\eta^2(3\eta - 5)(\eta-3) < 0 \quad \text{for} \quad 0 < \eta < \frac{5}{3}.$$

Therefore if $0 < B < \frac{5}{3}$ $u_1 \equiv 0$ supports two types of boundary layers at $t = 1$. Namely, if $0 < B \leq 1$ there is a U-layer and if $1 < B < \frac{5}{3}$ there is an S-layer. If $B > \frac{5}{3}$ then u_1 cannot support a boundary layer at $t = 1$. (If $B = \frac{5}{3}$ it can be shown that there is also an S-layer at $t = 1$.)

For $u_2(t) = t + \frac{3}{2}$ we proceed similarly. Since $h(0,y) > 0$ for all $y > \frac{3}{2}$ and $h(1,y) > 0$ for all $y > \frac{5}{2}$ it follows that

$$\int_{u_2(0)=\frac{3}{2}}^{\eta} h(0,s)\,ds > 0 \quad \text{for all} \quad \eta > \frac{3}{2} \quad \text{and}$$

$$\int_{u_2(1)=\frac{5}{2}}^{\eta} h(1,s)\,ds > 0 \quad \text{for all} \quad \eta > \frac{5}{2} \;.$$

Thus if $A > \frac{3}{2}$ $(B > \frac{5}{2})$ u_2 supports convex boundary layers at $t = 0$ $(t = 1)$. Suppose next that $A < \frac{3}{2}$ then for $\eta < \frac{3}{2}$

$$\pi_0(\eta) = \int_{\eta}^{u_2(0)=\frac{3}{2}} h(0,s)\,ds = -\frac{1}{12}(3\eta^4 - 10\eta^3 + 9\eta^2 - \frac{27}{16})$$

$$= -\frac{1}{24}(\eta - \frac{3}{2})^2(12\eta^2 - 4\eta - 3) \quad ,$$

and so $\pi_0(\eta) < 0$ for $\frac{1}{6}(1 + \sqrt{10} < \eta < \frac{3}{2}$. Thus for $\frac{1}{6}(1 + \sqrt{10})$ $\leq A < \frac{3}{2}$ u_2 supports a $\cap$-layer (if $1 \leq A < \frac{3}{2}$) or an S-layer (if $\frac{1}{6}(1 + \sqrt{10}) \leq A < 1$) at $t = 0$. It remains to examine boundary layer phenomena for $B < \frac{5}{2}$, namely to consider

$$\pi_1(\eta) = \int_{\eta}^{u_2(1)=\frac{5}{2}} h(1,s)\,ds = -\frac{1}{12}(3\eta^4 - 14\eta^3 + 15\eta^2 + \frac{375}{48})$$

$$= -\frac{1}{48}(\eta - \frac{5}{2})^2\,(12\eta^2 + 4\eta + 5) \quad .$$

Since $12\eta^2 + 4\eta + 5 > 0$ for all η it follows that $\pi_1(\eta) < 0$ for all $\eta < \frac{5}{2}$. Consequently u_2 supports three types of boundary layers at $t = 1$. Namely, there is a $\cap$-layer if $1 \leq B < \frac{5}{2}$, a Z-layer if $0 \leq B < 1$, and a concave-convex-concave boundary layer if $B < 0$; see figure 13.

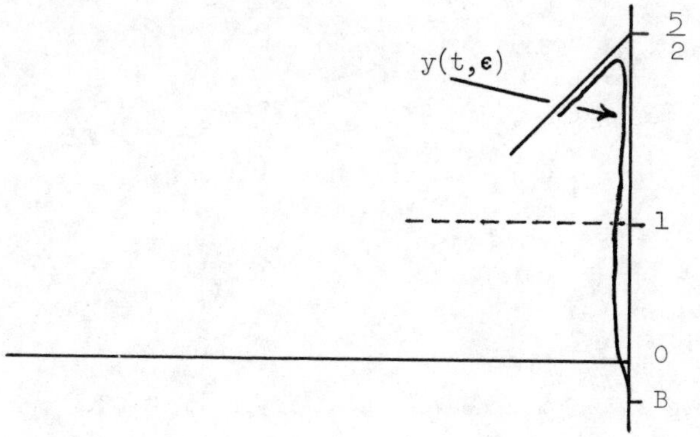

Fig. 13

Concave-Convex-Concave Boundary Layer at $t = b$.

Consider next the occurrence of shock layers by looking at

$$J[t] = \int_{u_1=0}^{u_2=t+\frac{3}{2}} h(t,s)\,ds \quad .$$

One easily sees that $t^* = \frac{1}{2}$ is the only zero of J and that

$J'[t] < 0$ for $0 \le t \le 1$. We conclude that all shock layers occur at $t^* = \frac{1}{2}$ and that these shocks are S-shocks.

We can now discuss the existence and the asymptotic behavior of solutions of (E1) for all values of A and B. All of our results follow from the corresponding theory developed above. Let us fix a value of B and see how solutions of (E1) behave for ϵ small and positive as A varies.

(I) $B > \frac{5}{2}$.

a) $A > \frac{3}{2}$. Since $u_2(0) < A$ and $u_2(1) < B$ (E1) has a solution $y_1 = y_1(t,\epsilon)$ such that $\lim_{\epsilon\to 0^+} y_1(t,\epsilon) = u_2(t) = t + \frac{3}{2}$ for $0 < t < 1$. In addition, there is another solution $y_2 = y_2(t,\epsilon)$ such that

$$\lim_{\epsilon\to 0^+} y_2(t,\epsilon) = \begin{cases} 0, & 0 < t < \frac{1}{2}, \\ t + \frac{3}{2}, & \frac{1}{2} < t < 1, \end{cases}$$

since $u_1 \equiv 0$ supports a convex-concave-convex layer at $t = 0$ and u_1, u_2 support an S-shock at $t = \frac{1}{2}$.

b) $A = \frac{3}{2}$. Since $u_2(0) = A$, $\lim_{\epsilon\to 0^+} y_1(t,\epsilon) = t + \frac{3}{2}$ for $0 \le t < 1$ while

$$\lim_{\epsilon\to 0^+} y_2(t,\epsilon) = \begin{cases} 0, & 0 < t < \frac{1}{2}, \\ t + \frac{3}{2}, & \frac{1}{2} < t < 1, \end{cases}$$

with a Z-boundary layer at $t = 0$.

c) $\frac{1}{6}(1+\sqrt{10}) \le A < \frac{3}{2}$. Here $u_2(0) > A$ and u_2 supports a $\cap$-layer (if $A \ge 1$) or an S-layer (if $A < 1$) at $t = 0$, while u_2 supports a $\cup$-layer at $t = 1$. Consequently (E1) has a solution $y_1 = y_1(t,\epsilon)$ such that $\lim_{\epsilon\to 0^+} y_1(t,\epsilon) = t + \frac{3}{2}$ for $0 < t < 1$. In addition, $u_1 \equiv 0$ supports a Z-layer at $t = 0$ (if $A \ge 1$) or a $\cup$-layer at $t = 0$ (if $A < 1$) and so (E1) has another solution $y_2 = y_2(t,\epsilon)$ such that

$$\lim_{\epsilon\to 0^+} y_2(t,\epsilon) = \begin{cases} 0 \,, & 0 < t < \frac{1}{2} \,, \\ t + \frac{3}{2} \,, & \frac{1}{2} < t < 1 \,. \end{cases}$$

d) $0 < A < \frac{1}{6}(1+\sqrt{10})$. For such A u_2 ceases to support a boundary layer at $t = 0$; however, $u_1 \equiv 0$ supports a $\cup$-layer at $t = 0$ and so (E1) has only the solution $y_2 = y_2(t,\epsilon)$ such that

$$\lim_{\epsilon\to 0^+} y_2(t,\epsilon) = \begin{cases} 0 \,, & 0 < t < \frac{1}{2} \,, \\ t + \frac{3}{2} \,, & \frac{1}{2} < t < 1 \,. \end{cases}$$

e) $A = 0$. Since $u_1(0) = A$

$$\lim_{\epsilon\to 0^+} y_2(t,\epsilon) = \begin{cases} 0 \,, & 0 \le t < \frac{1}{2} \,, \\ t + \frac{3}{2} \,, & \frac{1}{2} < t < 1 \,, \end{cases}$$

that is, there is no boundary layer at $t = 0$.

f) $A < 0$. For such A u_1 supports a $\cap$-boundary layer at $t = 0$ and so

$$\lim_{\epsilon\to 0^+} y_2(t,\epsilon) = \begin{cases} 0\ , & 0 < t < \frac{1}{2}\ , \\ t + \frac{3}{2}\ , & \frac{1}{2} < t < 1\ . \end{cases}$$

(II) $B = \frac{5}{2}$. The results in (a)-(f) of (I) are valid without a boundary layer at $t = 1$ because $u_2(1) = B$.

(III) $\frac{5}{3} \leq B < \frac{5}{2}$. For such B the results in (a)-(f) of (I) are valid with the exception that the solutions of (E1) have $\cap$-boundary layers at $t = 1$. This follows because $h(1,y) < 0$ for $\frac{5}{3} \leq y < \frac{5}{2}$.

(IV) $0 < B < \frac{5}{3}$.

a) $A > \frac{3}{2}$. Since $u_2(0) < A$ and since u_2 supports a Z-layer at $t = 1$ (E1) has a solution $y_1 = y_1(t,\epsilon)$ such that $\lim_{\epsilon\to 0^+} y_1(t,\epsilon) = t + \frac{3}{2}$ for $0 < t < 1$. Also, $u_1 \equiv 0$ supports a convex-concave-convex boundary layer at $t = 0$ and so (E1) has a second solution $y_2 = y_2(t,\epsilon)$ such that

$$\lim_{\epsilon\to 0^+} y_2(t,\epsilon) = \begin{cases} 0\ , & 0 < t < \frac{1}{2}\ , \\ t + \frac{3}{2}\ , & \frac{1}{2} < t < 1\ . \end{cases}$$

In addition, for such B u_1 supports an S-layer at $t = 1$ (if $B > 1$) or a U-layer there (if $B \leq 1$) and so (E1) has a third solution $y_3 = y_3(t,\epsilon)$ such that $\lim_{\epsilon\to 0^+} y_3(t,\epsilon) = 0$ for $0 < t < 1$.

b) $A = \frac{3}{2}$. Since $u_2(0) = A$, $\lim_{\epsilon\to 0^+} y_1(t,\epsilon) = t + \frac{3}{2}$ for $0 \leq t < 1$ while

$$\lim_{\epsilon\to 0^+} y_2(t,\epsilon) = \begin{cases} 0\ , & 0 < t < \frac{1}{2}\ , \\ t + \frac{3}{2}\ , & \frac{1}{2} < t < 1\ , \end{cases}$$

and $\lim_{\epsilon\to 0^+} y_3(t,\epsilon) = 0$ for $0 < t < 1$, both with Z-layers at $t = 0$.

c) $\frac{1}{6}(1 + \sqrt{10}) \le A < \frac{3}{2}$. Clearly $\lim_{\epsilon\to 0^+} y_1(t,\epsilon) = t + \frac{3}{2}$ for $0 < t < 1$ with an S-boundary layer at $t = 0$ and a Z-boundary layer at $t = 1$. Similarly

$$\lim_{\epsilon\to 0^+} y_2(t,\epsilon) = \begin{cases} 0\ , & 0 < t < \frac{1}{2}\ , \\ t + \frac{3}{2}\ , & \frac{1}{2} < t < 1\ , \end{cases}$$

and $\lim_{\epsilon\to 0^+} y_3(t,\epsilon) = 0$ for $0 < t < 1$, with a Z-layer at $t = 0$ (if $A > 1$) or with a U-layer there (if $A \le 1$).

d) $0 < A < \frac{1}{6}(1 + \sqrt{10})$. For such A the y_1-solution disappears because u_2 cannot support a boundary layer at $t = 0$. However,

$$\lim_{\epsilon\to 0^+} y_2(t,\epsilon) = \begin{cases} 0\ , & 0 < t < \frac{1}{2}\ , \\ t + \frac{3}{2}\ , & \frac{1}{2} < t < 1\ , \end{cases}$$

and $\lim_{\epsilon\to 0^+} y_3(t,\epsilon) = 0$ for $0 < t < 1$, both with U-boundary layers at $t = 0$.

e) $A = 0$. Since $u_1(0) = A$

$$\lim_{\epsilon\to 0^+} y_2(t,\epsilon) = \begin{cases} 0, & 0 \le t < \frac{1}{2}, \\ t + \frac{3}{2}, & \frac{1}{2} < t < 1, \end{cases}$$

and $\lim_{\epsilon\to 0^+} y_3(t,\epsilon) = 0$ for $0 \le t < 1$.

f) $A < 0$. Since $u_1 \equiv 0$ supports a $\cap$-layer at $t = 0$,

$$\lim_{\epsilon\to 0^+} y_2(t,\epsilon) = \begin{cases} 0, & 0 < t < \frac{1}{2}, \\ t + \frac{3}{2}, & \frac{1}{2} < t < 1, \end{cases}$$

and $\lim_{\epsilon\to 0^+} y_3(t,\epsilon) = 0$ for $0 < t < 1$.

(V) $B = 0$. Since $u_1(1) = B$ the results in (a)-(f) of (IV) are valid with no boundary layer at $t = 1$.

(VI) $B < 0$. For such B the results in (a)-(f) of (IV) are valid with the exception that the solutions of (E1) have concave-convex-concave boundary layers at $t = 1$.

This concludes our discussion of (E1) . We note first that the behavior of the y_2-solution in (c) and (d) of case (III) illustrates the interaction between boundary and shock layer behavior described at the end of Section 4. Secondly the values $A^* = \frac{1}{6}(1 + \sqrt{10})$ and $B^* = \frac{5}{3}$ act like bifurcation parameters since the number of solutions of (E1) changes as A crosses A^* and B crosses B^* . For the sake of clarity we give a graphical representation of the solution set of (E1) in the case that $0 < B < \frac{5}{3}$; see figures 14-19.

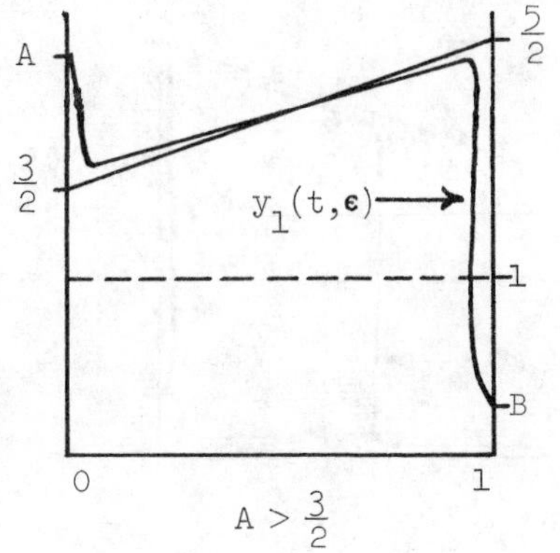

Fig. 14(i)

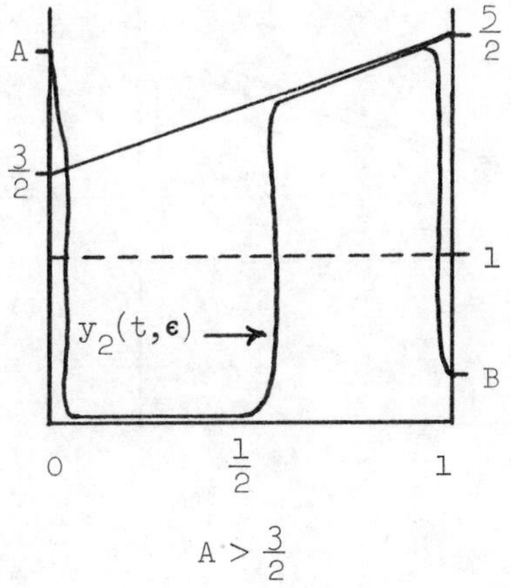

Fig. 14(ii)

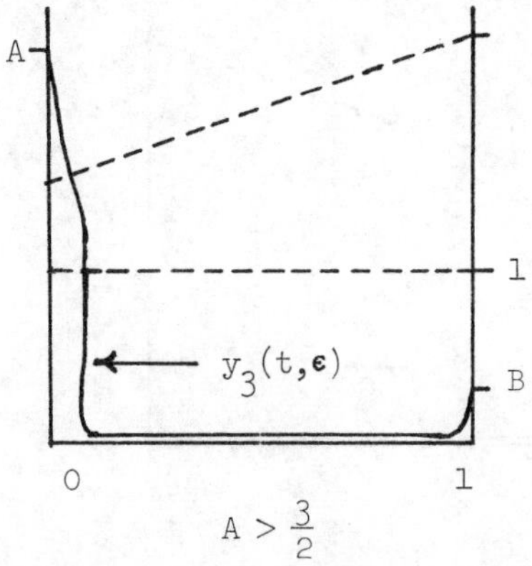

Fig. 14(iii)

The Three Solutions of (E1) for $A > \frac{3}{2}$.

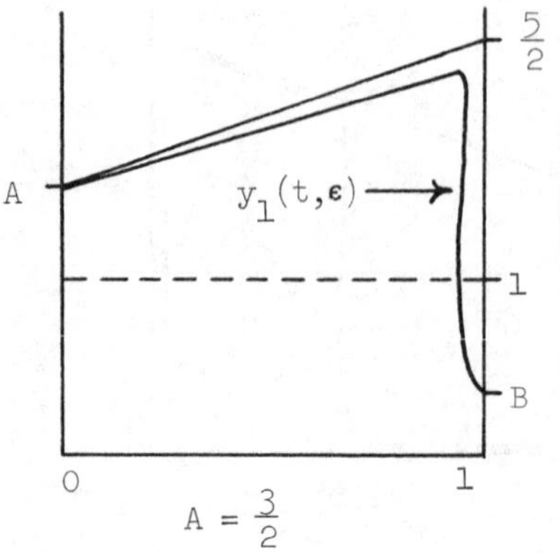

Fig. 15 (i)

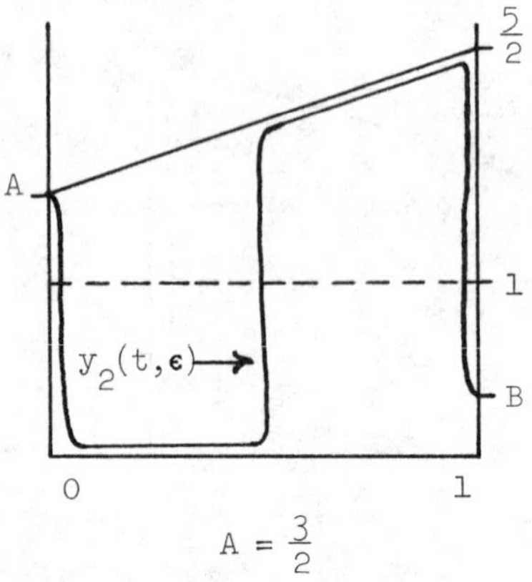

Fig. 15 (ii)

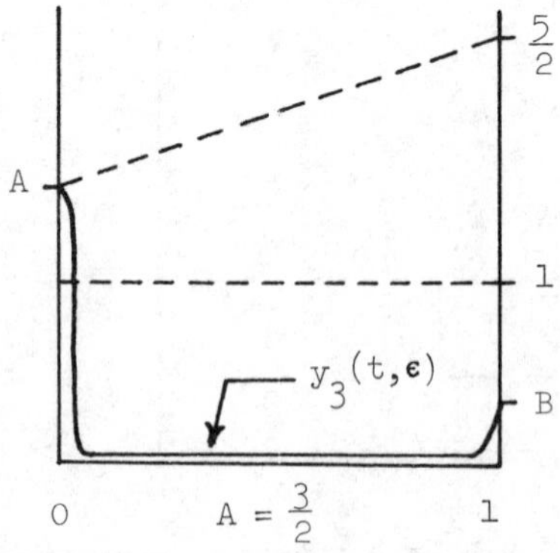

Fig. 15(iii)

The Three Solutions of (E1) for $A = \frac{3}{2}$.

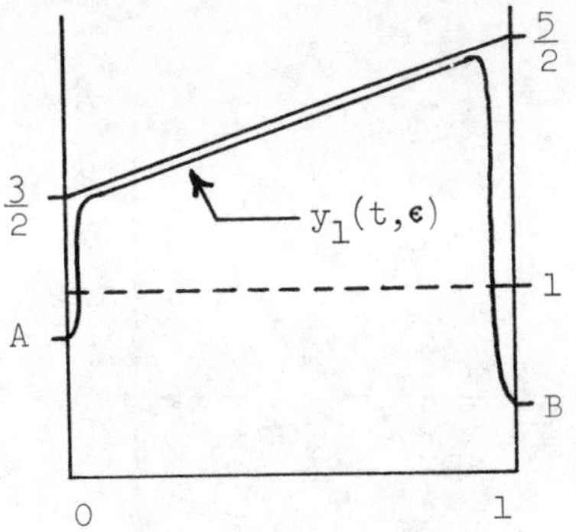

$\frac{1}{6}(1+\sqrt{10}) < A < \frac{3}{2}$

Fig. 16 (i)

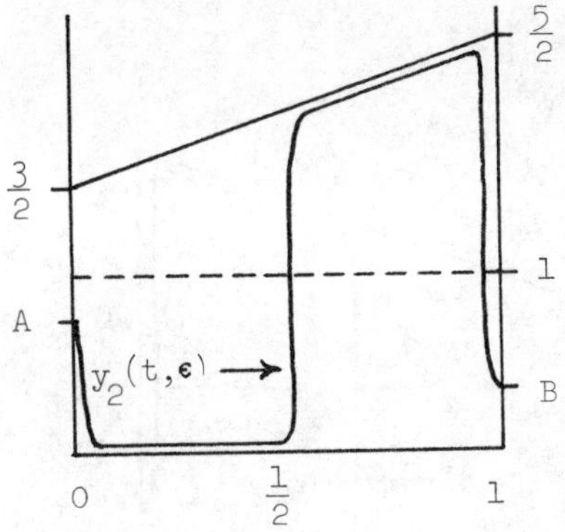

$\frac{1}{6}(1+\sqrt{10}) < A < \frac{3}{2}$

Fig. 16 (ii)

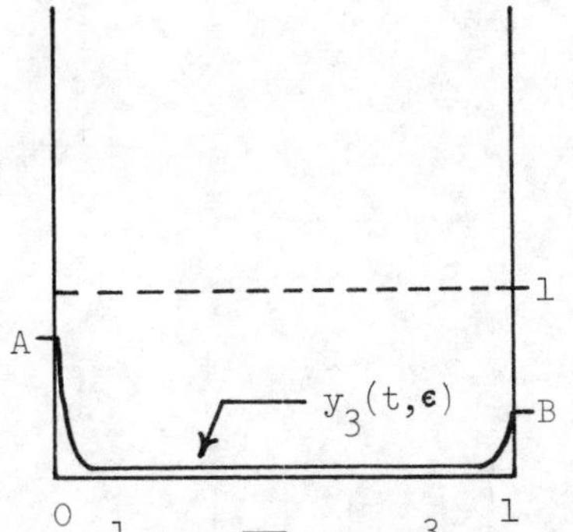

$\frac{1}{6}(1+\sqrt{10}) < A < \frac{3}{2}$

Fig. 16 (iii)

The Three Solutions of (E1) for $\frac{1}{6}(1+\sqrt{10}) < A < \frac{3}{2}$.

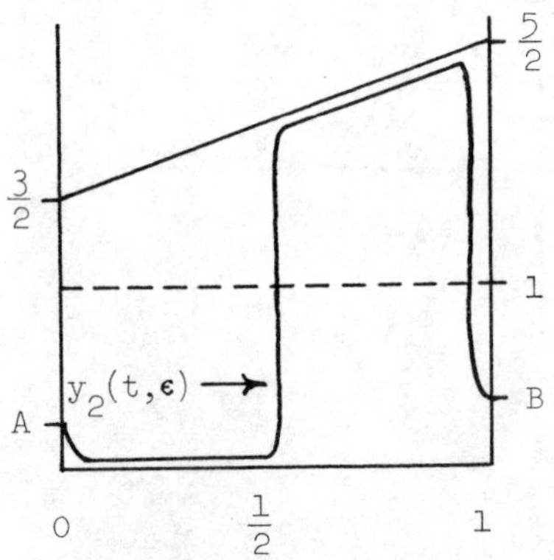

$0 < A \leq \frac{1}{6}(1+\sqrt{10})$

Fig. 17 (i)

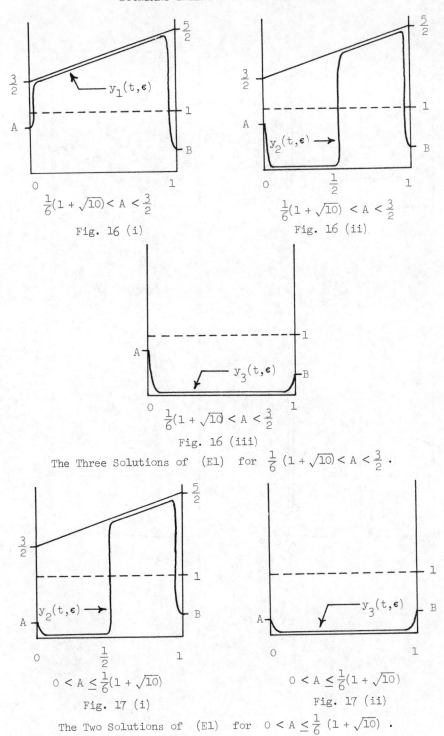

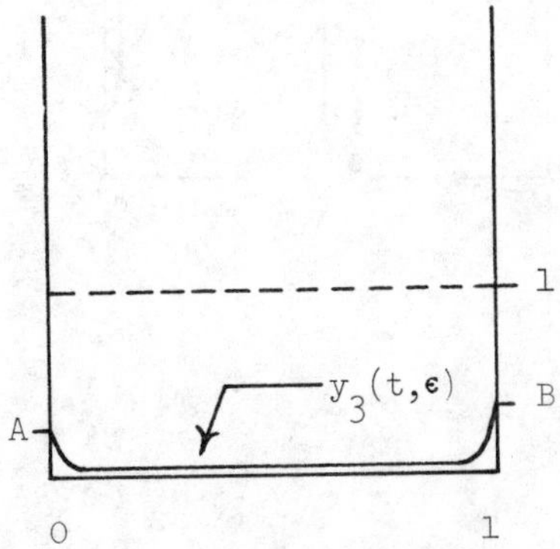

$0 < A \leq \frac{1}{6}(1+\sqrt{10})$

Fig. 17 (ii)

The Two Solutions of (E1) for $0 < A \leq \frac{1}{6}(1+\sqrt{10})$.

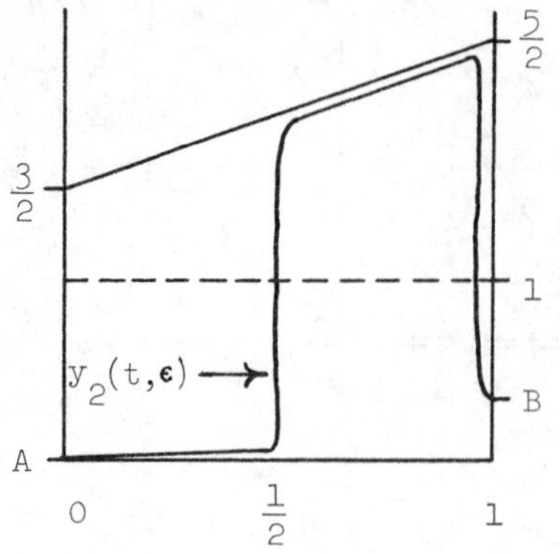

A = 0

Fig. 18 (i)

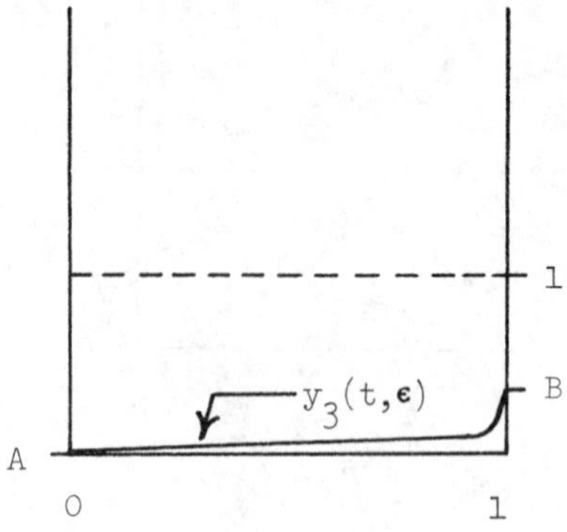

A = 0

Fig. 18 (ii)

The Two Solutions of (E1) for $A = 0$.

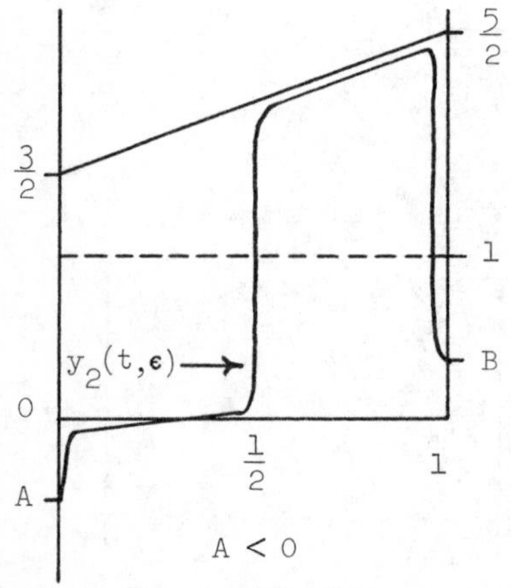

Fig. 19 (i)

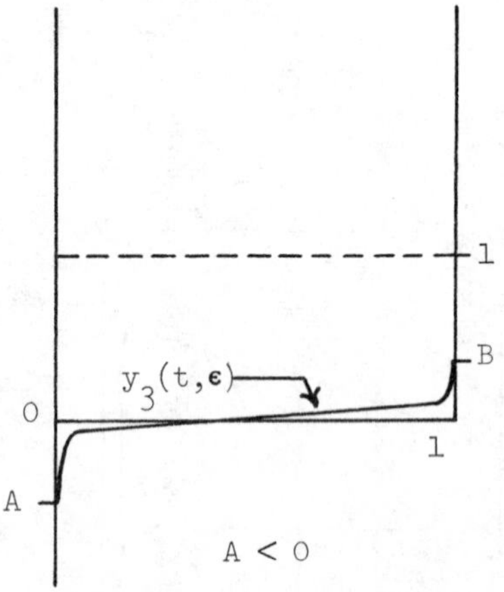

Fig. 19 (ii)

The Two Solutions of (E1) for $A < 0$.

We consider next the problem

$$\epsilon y'' = y(y-1)(y-(\tfrac{1}{2}t+\tfrac{1}{4})) = h(t,y)\ , \quad 0 < t < 1\ ,$$

(E2)

$$y(0,\epsilon) = A\ , \quad y(1,\epsilon) = B\ .$$

Proceeding as in the discussion of (E1) it follows that $u_1 \equiv 0$ and $u_2 \equiv 1$ are $(I)_0$-stable reduced solutions while $\tilde{u}(t) = \frac{1}{2}t + \frac{1}{4}$ is unstable. Short computations show that

$$\int_0^\eta h(0,s)ds > 0 \quad \text{for} \quad 0 < \eta < \tfrac{1}{6}(5-\sqrt{7})\ ,$$

$$\int_\eta^0 h(0,s)ds < 0 \quad \text{for} \quad \eta < 0\ ;$$

$$\int_0^\eta h(1,s)ds > 0 \quad \text{for} \quad \eta > 0\ ,$$

$$\int_\eta^0 h(1,s)ds < 0 \quad \text{for} \quad \eta < 0\ ;$$

$$\int_1^\eta h(0,s)ds > 0 \quad \text{for} \quad 1 < \eta\ ,$$

$$\int_\eta^1 h(0,s)ds < 0 \quad \text{for} \quad \eta < 1\ ;$$

$$\int_1^\eta h(1,s)ds > 0 \quad \text{for} \quad 1 < \eta\ ,$$

$$\int_\eta^1 h(1,s)ds < 0 \quad \text{for} \quad \tfrac{1}{6}(1+\sqrt{7}) < \eta < 1\ .$$

Consequently at $t = 0$ $u_1 \equiv 0$ supports $\cap$ - $(A < 0)$, $\cup$ - $(0 < A \leq \frac{1}{4})$ and Z- $(\frac{1}{4} < A < \frac{1}{6}(5-\sqrt{7}))$ boundary layers, while at $t = 1$ u_1 supports $\cap$-$(B < 0)$, $\cup$-$(0 < B \leq \frac{3}{4})$, S-$(\frac{3}{4} < B \leq 1)$ and convex-concave-convex $(B > 1)$ boundary layers. Likewise at $t = 0$ $u_2 \equiv 1$

supports U-$(A > 1)$, $\cap$-$(\frac{1}{4} \le A < 1)$, S-$(0 \le A < \frac{1}{4})$ and concave-convex-concave $(A < 0)$ boundary layers, while at $t = 1$ u_2 supports U-$(B > 1)$, $\cap$-$(\frac{3}{4} \le B < 1)$ and Z-$(\frac{1}{6}(1 + \sqrt{7}) \le B < \frac{3}{4})$ boundary layers. If $A > \frac{1}{6}(5 - \sqrt{7})$ u_1 cannot support a boundary layer at $t = 0$ and if $B < \frac{1}{6}(1 + \sqrt{7})$ u_2 cannot support a boundary layer at $t = 1$.

To check for shock layer behavior consider

$$J[t] = \int_0^1 h(t,s)\,ds \quad .$$

Clearly $J'[t] > 0$ and $t^* = \frac{1}{2}$ is the only zero of J. We conclude that any shock layer occurs at $t^* = \frac{1}{2}$ and that all shocks are Z-shocks. Using these facts about the boundary and shock layer structure we can give a complete account of the existence and the asymptotic behavior of solutions of (E2) . These results are displayed graphically in the case $\frac{1}{6}(1 + \sqrt{7}) < B < 1$ in figures 20-25.

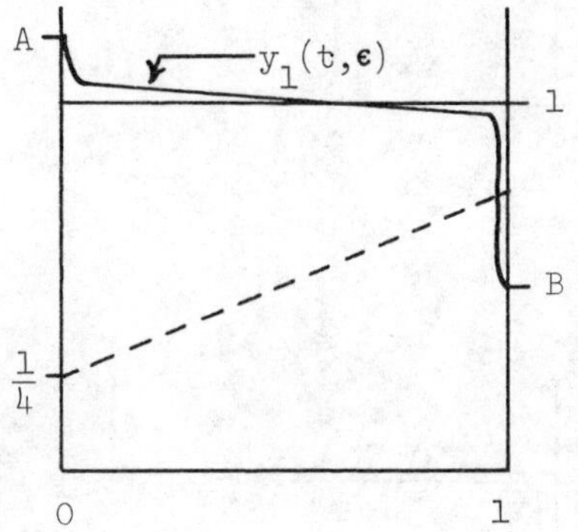

$A > 1$

Fig. 20 (i)

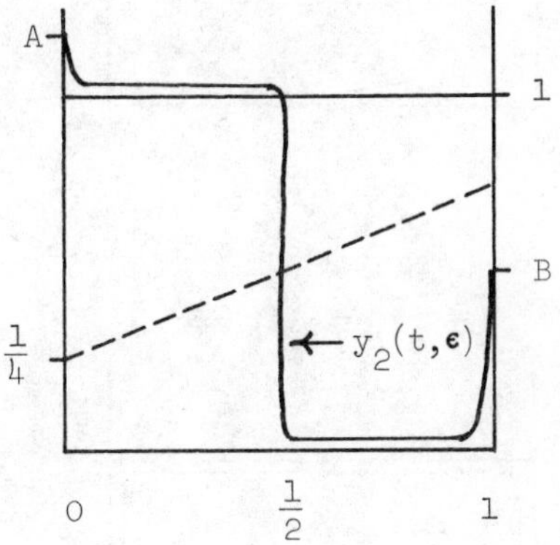

$A > 1$

Fig. 20 (ii)

The Two Solutions of (E2) for $A > 1$.

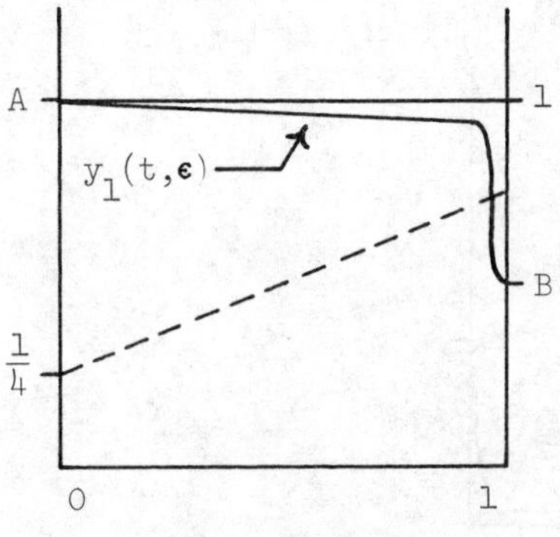

$A = 1$

Fig. 21 (i)

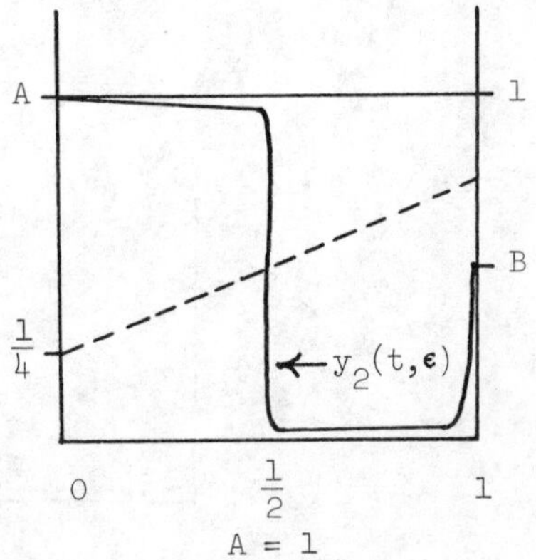

$A = 1$

Fig. 21 (ii)

The Two Solutions of (E2) for $A = 1$.

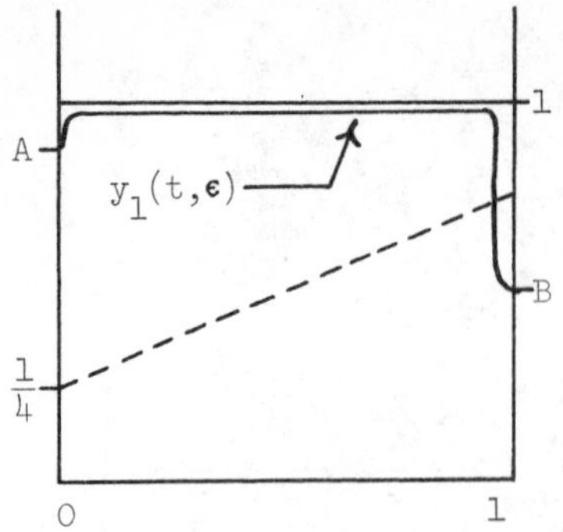

$\frac{1}{6}(5 - \sqrt{7}) \leq A < 1$

Fig. 22 (i)

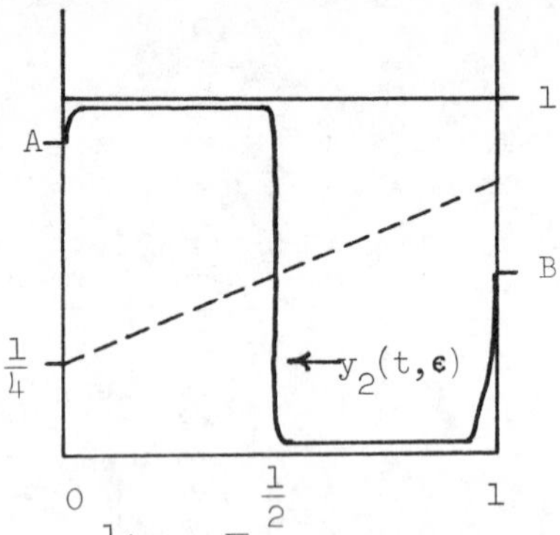

$\frac{1}{6}(5 - \sqrt{7}) \leq A < 1$

Fig. 22 (ii)

The Two Solutions of (E2) for $\frac{1}{6}(5 - \sqrt{7}) \leq A < 1$.

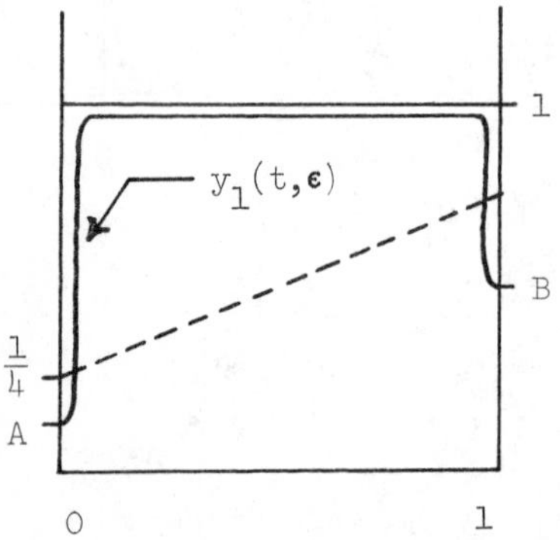

$0 < A < \frac{1}{6}(5 - \sqrt{7})$

Fig. 23 (i)

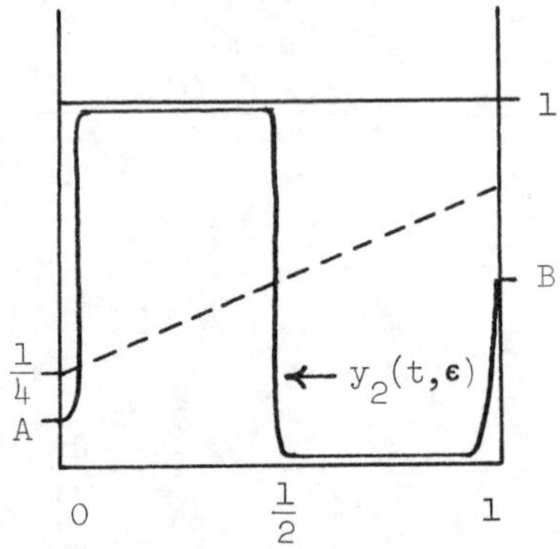

$0 < A < \frac{1}{6}(5 - \sqrt{7})$

Fig. 23 (ii)

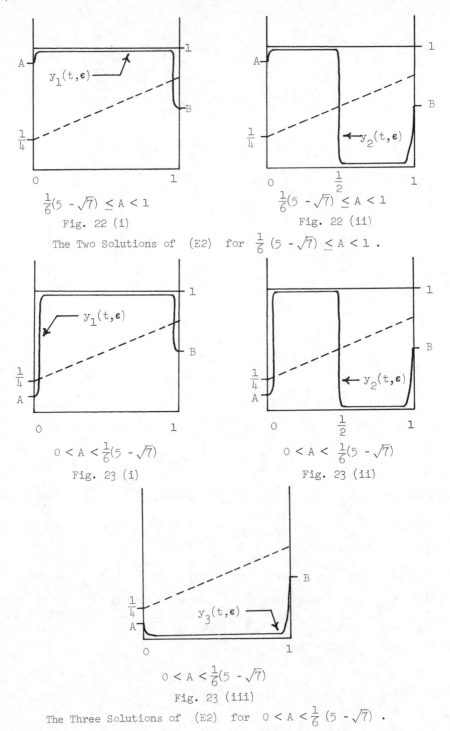

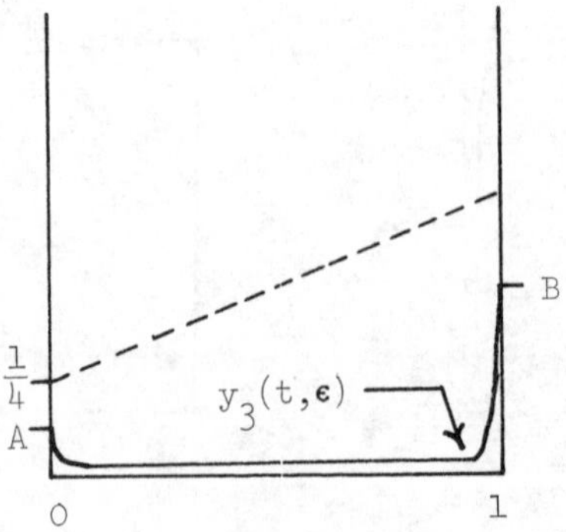

$0 < A < \frac{1}{6}(5 - \sqrt{7})$

Fig. 23 (iii)

The Three Solutions of (E2) for $0 < A < \frac{1}{6}(5 - \sqrt{7})$.

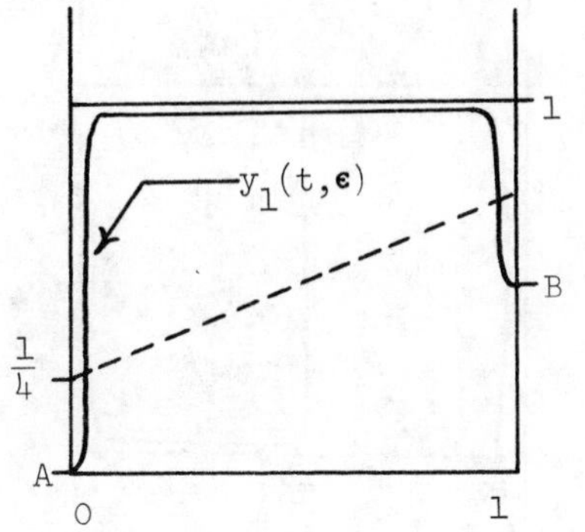

A = 0

Fig. 24 (i)

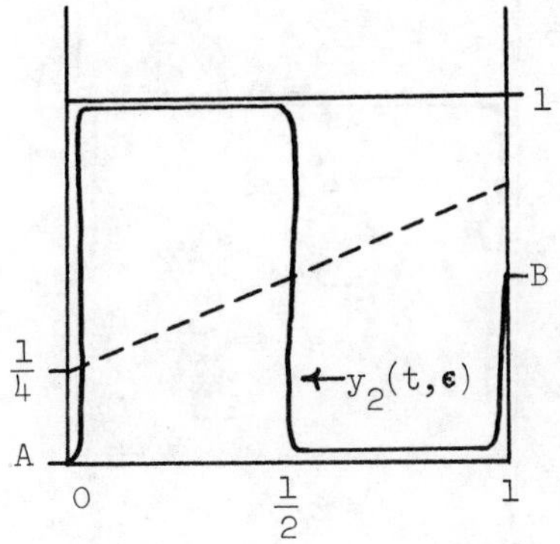

A = 0

Fig. 24 (ii)

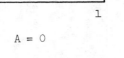

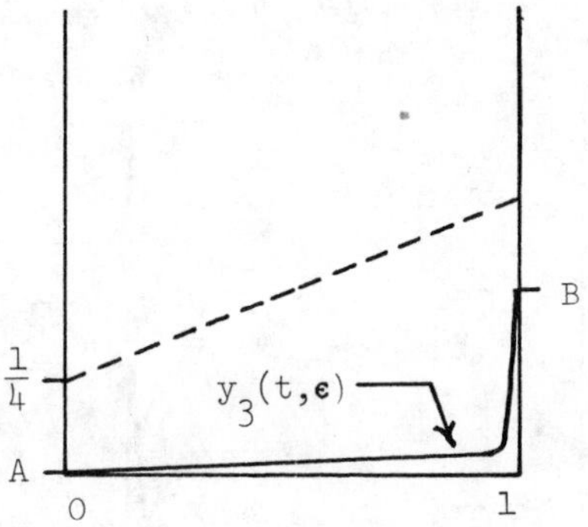

A = 0

Fig. 24 (iii)

The Three Solutions of (E2) for A = 0 .

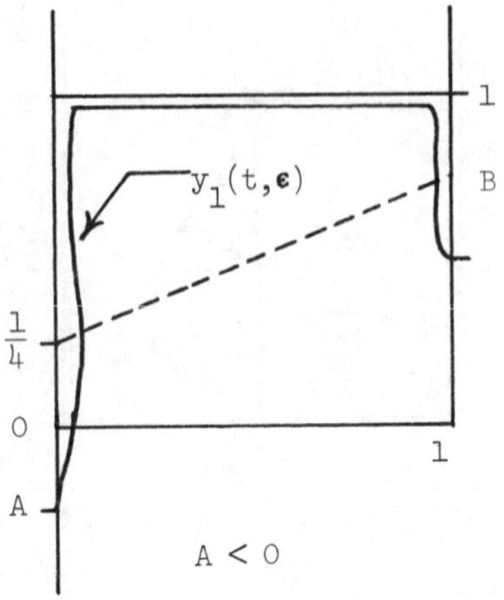

Fig. 25 (i)

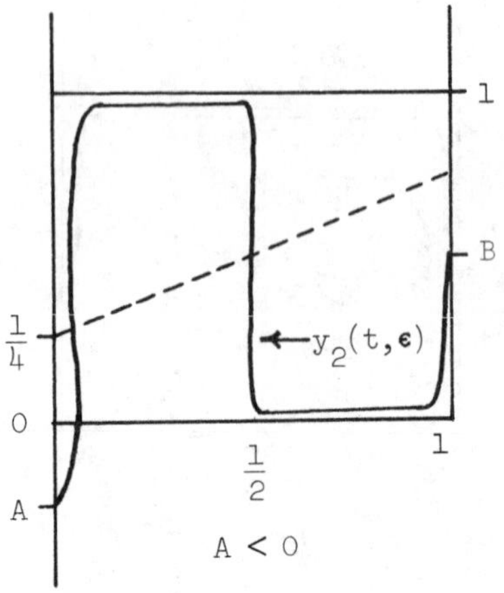

Fig. 25 (ii)

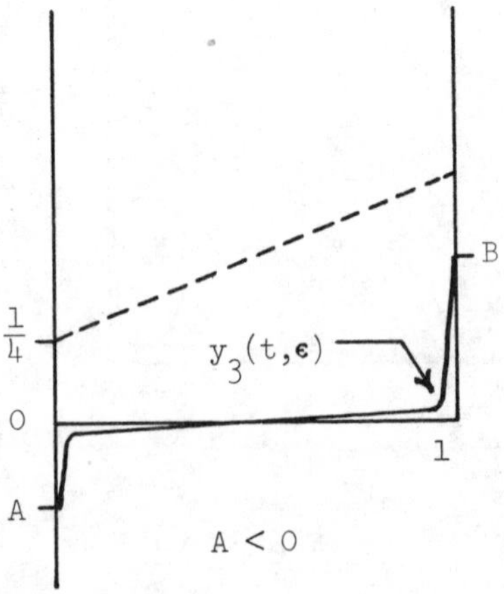

Fig. 25 (iii)

The Three Solutions of (E2) for $A < 0$.

8. Examples of (P_2) . We consider first a model problem which has interested many writers on singular perturbations (cf., for example, [6; pp. 29-38] and [7; Sec. 4]); namely,

$$\epsilon y'' = -yy' + y = f(t,y)y' + g(t,y)\ , \quad 0 < t < 1\ ,$$

(E3)

$$y(0,\epsilon) = A\ , \quad y(1,\epsilon) = B\ .$$

The reduced equation has the regular solutions $u_L(t) = t + A$, $u_R(t) = t + B - 1$ and the singular solution $u_s \equiv 0$. Since $f(t,y) = -y$ the solution u_L is globally stable if $A \leq -1$, while u_R is globally stable if $B \geq 1$. The solution u_s is $(I)_0$-stable since $g_y \equiv 1$. Moreover, u_L is partially stable on $[0,-A)$ for $-1 < A < 0$, that is, $t_L = -A$, and u_R is partially stable on $(1 - B, 1]$ for $0 < B < 1$, that is, $t_R = 1 - B$.

Consider first the case when u_R is globally stable, that is, $B \geq 1$, and let us examine for what values of A u_R supports a boundary layer at $t = 0$. It follows directly that for all $A > B-1$

$$\int_{u_R(0)=B-1}^{\eta} (-s)\,ds < 0 \quad \text{for} \quad B-1 < \eta \leq A$$

and so u_R supports a U-layer at $t = 0$. Similarly

$$\int_{\eta}^{u_R(0)=B-1} (-s)\,ds < 0 \quad \text{for} \quad A \leq \eta < B-1$$

provided $|A| < B-1$, that is, $-(B-1) < A < B-1$. If $A \geq 0$ then u_R supports a ∩-layer at $t = 0$, while if $A < 0$ then the layer is of S-type. For $A = -(B-1)(B > 1)$ it is also possible to show that u_R supports an S-layer at $t = 0$.

Consider next the case when u_L is globally stable, that is, $A \leq -1$, and suppose first that $B < A+1$. It is clear that

$$\int_{\eta}^{u_L(1)=A+1} (-s)\,ds > 0 \quad \text{for} \quad B \leq \eta < A+1$$

for all $B < A+1$, that is, u_L supports a $\cap$-boundary layer at $t = 1$. If however $B > A+1$ then

$$\int_{u_L(1)=A+1}^{\eta} (-s)\,ds > 0 \quad \text{for} \quad A+1 < \eta \leq B \quad ,$$

provided $|B| < -(A+1)$, that is, $A+1 < B < -(A+1)$. Thus for $B \leq 0$ u_L supports a $\cup$-layer at $t = 1$ and for $B > 0$ there is a boundary layer of S-type there. If $B = -(A+1)\,(A < -1)$ then u_L also supports an S-layer at $t = 1$.

Finally if $B = 1$ then u_R supports either a $\cup$-layer at $t = 0$ (if $A > 0$) or a $\cap$-layer there (if $A < 0$), while if $A = -1$ then u_L supports either a $\cap$-layer at $t = 1$ (if $B < 0$) or a $\cup$-layer there (if $B > 0$).

From our observations in the latter half of Section 5 we know that if simultaneously u_L cannot support an S-layer at $t = 1$ (of course we still assume $u_L(1) < B$ and $u_R(0) > A$) and u_R cannot support an S-layer at $t = 0$ then $J[t^*] = 0$ at a point t^* in $(0,1)$ for

$$J[t] = \int_{t+A}^{t+B-1} (-s)\,ds \quad .$$

(Moreover, $J'[t] < 0$ for $0 \leq t \leq 1$.) In other words u_L and u_R support an S-shock layer at the point t^* which is easily seen to have the value $t^* = \frac{1}{2}(1-B-A)$. Furthermore, as noted in Section 5, there is also an S-shock layer at this t^* if u_L is globally stable

$(A \leq -1)$ and u_R is only stable on $(t_R,1](t_R = 1-B$ in $(0,1))$ or if u_R is globally stable $(B \geq 1)$ and u_L is only stable on $[0,t_L)$ $(t_L = -A$ in $(0,1))$. In the first case t^* clearly belongs to $(t_R,1)$ and in the second t^* belongs to $(0,t_L)$.

Next suppose that u_L and u_R are both partially stable, that is, $0 < t_L, t_R < 1$. If $t_L < t_R$ then the domains of stability do not overlap and $u_L(t_L) = u_R(t_R) = u_s \equiv 0$. Since u_s is stable we know that (cf. Theorem 5.6) (E3) has a solution $y = y(t,\epsilon)$ such that

$$\lim_{\epsilon\to 0^+} y(t,\epsilon) = \begin{cases} t+A, & 0 \leq t \leq -A, \\ 0, & -A \leq t \leq 1-B, \\ t+B-1, & 1-B \leq t \leq 1. \end{cases}$$

The case $t_L = t_R$ is trivial since $u_L \equiv u_R$; indeed, $y(t,\epsilon) = t+A = t+B-1$ is an exact solution of (E3). Finally suppose that $t_L > t_R$ then our reasoning in Section 5 lets us conclude that u_L and u_R support an S-shock layer at $t^* = \frac{1}{2}(1-B-A)$ in (t_R,t_L).

In looking over our discussion we see that three cases have yet to be treated; namely, (i) $A > 0$ and $B < 0$ (ii) $0 < B < 1$ and $A \geq 0$ (iii) $-1 < A < 0$ and $B \leq 0$. The asymptotic behavior here is described with the aid of the stable singular solution $u_s \equiv 0$. In case (i) u_s is compatible with the boundary conditions and so by Theorem 5.7 we deduce the existence of a solution $y = y(t,\epsilon)$ of (E3) such that

$$B \exp[-(1-t)\epsilon^{-1/2}] \leq y(t,\epsilon) \leq A \exp[-t\epsilon^{-1/2}] \quad \text{for} \quad 0 \leq t \leq 1 .$$

In case (ii) u_R loses stability at t_R in $(0,1)$, that is, u_R crosses u_s at t_R and so (E3) has a solution $y = y(t,\epsilon)$ such that

$$\lim_{\epsilon\to 0^+} y(t,\epsilon) = \begin{cases} 0\ , & 0 < t \le 1-B\ , \\ t+B-1\ , & 1-B \le t \le 1\ , \end{cases}$$

(cf. our remarks at the end of Section 5). Finally in case (iii) u_L loses stability at t_L in $(0,1)$, that is, $u_L(t_L) = 0$ and so (E3) has a solution $y = y(t,\epsilon)$ such that

$$\lim_{\epsilon\to 0^+} y(t,\epsilon) = \begin{cases} t+A\ , & 0 \le t \le -A\ , \\ 0\ , & -A \le t < 1\ . \end{cases}$$

It follows from this discussion that at least for $\epsilon > 0$ sufficiently small the problem (E3) has a unique solution for all values of the boundary conditions A and B (cf. also [7 ; Sec. 4].).

We summarize our results in the form of a boundary value portrait (cf. [6; p. 33]) figure 26.

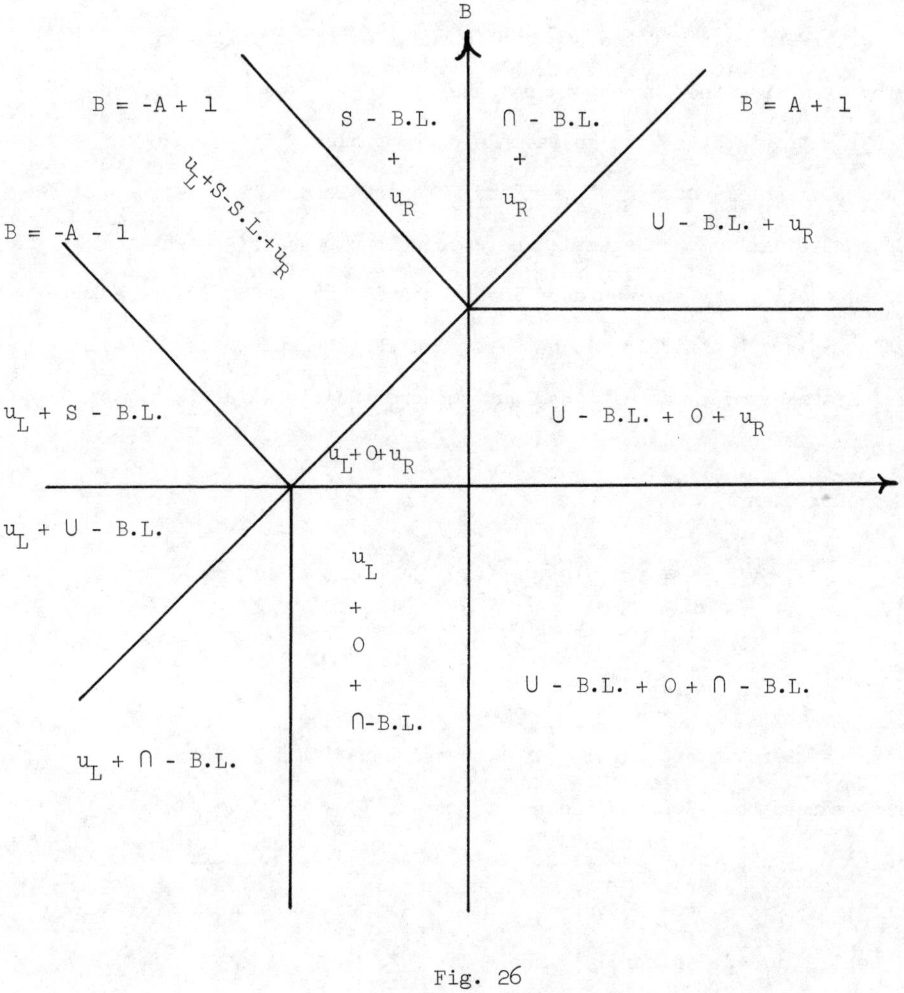

Fig. 26

Boundary Value Portrait of (E3) ;
B.L. (S.L.) denotes boundary (shock) layer .

If now we make the change of dependent variable $y \to -y$ in (E3) then we obtain the problem

$$\text{(E3)}' \qquad \begin{aligned} &\epsilon y'' = yy' + y \ , \quad 0 < t < 1 \ , \\ &y(0,\epsilon) = A \ , \quad y(1,\epsilon) = B \ . \end{aligned}$$

It is possible to read off the asymptotic behavior of solutions of (E3)' by examining the corresponding case of (E3). We note however that (E3)' provides us with an example of a problem (P_2) whose solutions possess Z-boundary layers and Z-shock layers.

We turn finally to an example whose solutions possess somewhat unexpected oscillatory and boundary layer behavior. The theory presented in Section 5 is not entirely adequate to describe the solutions for all choices of the boundary conditions, and so some additional devices must be employed.

Consider then

$$\text{(E4)} \qquad \begin{aligned} &\epsilon y'' = yy' - y = f(t,y)y' + g(t,y) \ , \quad 0 < t < 1 \ , \\ &y(0,\epsilon) = A \ , \quad y(1,\epsilon) = B \ , \end{aligned}$$

which is the steady-state version of Burgers' equation [2], [26].

The reduced equation has the solutions $u_L(t) = t+A$, $u_R(t) = t+B-1$ and $u_s \equiv 0$; in addition, $u_I(t) = t+C$ is a solution for any constant C. Since $f(t,y) = y$ u_L is (globally) stable if $A \geq 0$, while u_R is (globally) stable if $B \leq 0$. If $A < 0$ then u_L is unstable; however, u_L is stable with respect to boundary layer behavior at $t = 1$ if $A > -1$. Similarly if $B > 0$ u_R is unstable; however, it is stable with respect to boundary layer behavior at $t = 0$ if $B < 1$. The singular solution u_s is unstable because $g_y \equiv -1$. Finally u_I is stable with respect to boundary layer behavior at each endpoint if $-1 < C < 0$. Thus despite the fact that (E3) and (E4) have the same reduced solutions their stability properties are fundamentally different.

Suppose first that $B \leq 0$ and let us look for boundary layer behavior at $t = 0$. If $A < B-1$

$$\int_{\eta}^{u_R(0)=B-1} s\,ds < 0 \quad \text{for} \quad A \leq \eta < B-1$$

and so u_R supports a $\cap$-boundary layer at $t = 0$. However if $A > B-1$ $\int_{B-1}^{\eta} s\,ds < 0$ for $B-1 < \eta \leq A$ <u>provided</u> $|A| < -(B-1)$, that is, $B-1 < A < -(B-1)$. Thus u_R supports a $\cup$-layer (if $A \leq 0$) or a Z-layer (if $A > 0$) at $t = 0$.

Similarly if $A \geq 0$ it follows directly that for $B > A+1$

$$\int_{u_L(1)=A+1}^{\eta} s\,ds > 0 \quad \text{for} \quad A+1 < \eta \leq B$$

and so u_L supports a $\cup$-boundary layer at $t = 1$. If however $B < A+1$ then

$$\int_{\eta}^{A+1} s\,ds > 0 \quad \text{for} \quad B \leq \eta < A+1$$

<u>provided</u> $|B| < A+1$, that is, $-(A+1) < B < (A+1)$. Thus u_L supports a $\cap$-layer (if $B \geq 0$) or a Z-layer (if $B < 0$) at $t = 1$.

Suppose now that $A \geq 0$ and $B \leq 0$ and consider

$$J[t] = \int_{u_R(t)=t+B-1}^{u_L(t)=t+A} s\, ds \ .$$

Clearly $J[t] = 0$ only at $t^* = \frac{1}{2}$ and $J'[t] > 0$ for $0 \le t \le 1$ and so (E4) always has a solution $y_1 = y_1(t,\epsilon)$ such that

$$\lim_{\epsilon\to 0^+} y_1(t,\epsilon) = \begin{cases} t+A \ , & 0 \le t < \frac{1}{2} \ , \\ t+B-1 \ , & \frac{1}{2} < t \le 1 \ , \end{cases}$$

that is, u_L and u_R support a Z-shock layer at $t^* = \frac{1}{2}$. However if $A < -(B-1)$ we know from our previous calculation that u_R supports a Z-boundary layer at $t = 0$, that is, (E4) has another solution $y_2 = y_2(t,\epsilon)$ such that

$$\lim_{\epsilon\to 0^+} y_2(t,\epsilon) = t+B-1 \quad \text{for} \quad 0 < t \le 1 \ .$$

Similarly if $|B| < A+1$ we know that u_L supports a Z-boundary layer at $t = 1$ and so (E4) has a third solution $y_3 = y_3(t,\epsilon)$ such that

$$\lim_{\epsilon\to 0^+} y(t,\epsilon) = t+A \quad \text{for} \quad 0 \le t < 1 \ .$$

The behavior of the solutions of (E4) here provides an example of the interaction of boundary and shock layer behavior described in Section 5 for the case $J'[t] > 0$. However for such A and B (E4) has additional solutions whose existence is not predicted by the previous theory. Namely it can be shown (cf. [2]) that in addition to y_1, y_2, y_3 (E4) has solutions which exhibit boundary <u>and</u> shock layer behavior simultaneously and which cross the t-axis up to as many times as $O([\epsilon^{-1}])$ where $[\cdot]$ denotes the greatest integer function ; see figures 27 (i) and 27(ii) .

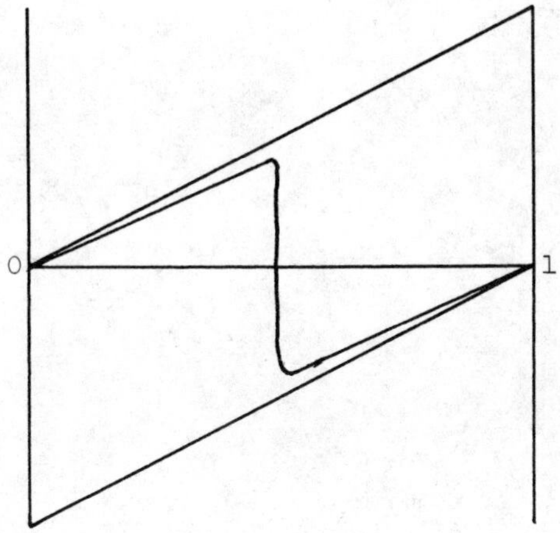

Fig. 27 (i)

Solution of (E4) for A = B = 0 having one Z-Shock.

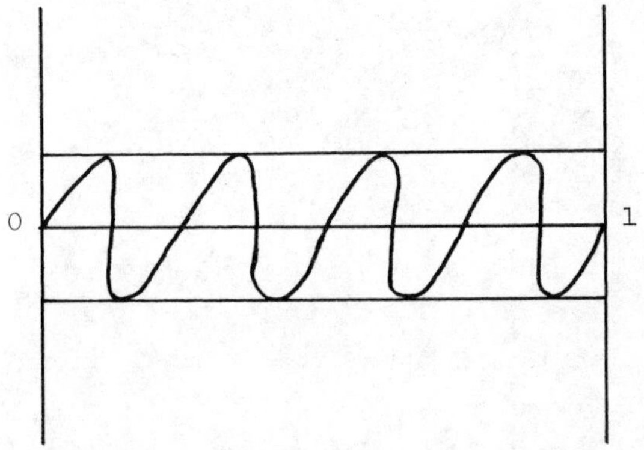

Fig. 27 (ii)

Solution of (E4) for A = B = 0 having four Z-Shocks .

Suppose finally that $A < 0$ and $B > 0$. The solutions u_L and u_R are now unstable along with $u_s \equiv 0$; however, we know by Nagumo's theorem that (E4) has a solution $y = y(t,\epsilon)$ such that for $0 \leq t \leq 1$

$$t + \min\{A,B-1\} \leq y(t,\epsilon) \leq t + \max\{A,B-1\} \quad .$$

Consider first the case $A < -1$ and $B > 1$; then in addition, neither u_L nor u_R is stable with respect to boundary layer behavior. It follows that

$$\lim_{\epsilon\to 0^+} y(t,\epsilon) = t+c \quad \text{for} \quad 0 < t < 1$$

for a constant c in $(-1,0)$, that is, u_I supports a $\cap$-layer at $t = 0$ and a $\cup$-layer at $t = 1$. To find c we begin by making the change of variable $w = y-t$ which transforms (E4) into

$$\epsilon w'' = (w+t)w' \quad , \quad 0 < t < 1 \quad ,$$
$$w(0,\epsilon) = A \quad , \quad w(1,\epsilon) = B-1 \quad .$$

Since $A < 0$ and $B-1 > 0$ $w'(t,\epsilon) > 0$ for $0 \leq t \leq 1$ and so the w-equation can be written as

$$\epsilon \frac{w''}{w'} = w + t \ , \quad \text{that is,} \quad \epsilon \frac{d}{dt}(\ell n\, w') = w + t \quad .$$

Integrating both sides of this equation from 0 to 1 gives

$$(\sim) \qquad \epsilon\, \ell n(w'(1,\epsilon)) - \epsilon\, \ell n(w'(0,\epsilon)) = \int_0^1 w(t,\epsilon)\,dt + \frac{1}{2} \quad .$$

Now the y-solution we are looking for corresponds to finding the solution $w = w(t,\epsilon)$ such that

$$w(t,\epsilon) \to c \quad \text{for} \quad 0 < t < 1 \ .$$

Since c supports boundary layers at $t = 0$ and $t = 1$ we know (cf. [34], [32; Chap. 2]) that $w'(1,\epsilon) = O(\epsilon^{-1})$ and $w'(0,\epsilon) = O(\epsilon^{-1})$, and so $\lim_{\epsilon\to 0^+} \{\epsilon\, \ell n(w'(1,\epsilon)) - \epsilon\, \ell n(w'(0,\epsilon))\} = 0$. Letting $\epsilon \to 0^+$ in (~) we have finally that $c = -\frac{1}{2}$ (since $\int_0^1 w(t,\epsilon)dt \to c$). In terms of (E4) this means that for $A < -1$ and $B > 1$ the solution $y = y(t,\epsilon)$ satisfies

$$\lim_{\epsilon\to 0^+} y(t,\epsilon) = t - \frac{1}{2} \quad \text{for} \quad 0 < t < 1 \ .$$

The remaining cases can be handled in a similar manner. Depending on the relative position of A and B (E4) has solutions which converge to $t+A$, $t+B-1$ and/or $t-\frac{1}{2}$ in $(0,1)$. A more general discussion of the problem $\epsilon y'' = h(t,y)y'$, $a < t < b$, $y(a,\epsilon)$, $y(b,\epsilon)$ prescribed, is contained in [24].

9. Examples of (P_3). Consider first the problem

$$\epsilon y'' = y - yy'^2 = F(t,y,y'), \quad 0 < t < 1 ,$$

(E5)

$$y(0,\epsilon) = A , \quad y(1,\epsilon) = B \ .$$

The reduced solutions to be considered are $u_L(t) = t+A$, $\tilde{u}_L(t) = A-t$, $u_R(t) = t+B-1$, $\tilde{u}_R(t) = B+1-t$ and the singular solution $u_s \equiv 0$. We begin by investigating their stability:

$$F_{y'}[t+A] = -2(t+A) \begin{cases} > 0 \text{ , if } A < -1 \text{ for } 0 \le t \le 1 \text{ ,} \\ > 0 \text{ , if } -1 < A < 0 \text{ for } 0 \le t < -A \text{ ;} \end{cases}$$

$$F_{y'}[A-t] = 2(A-t) \begin{cases} > 0 \text{ , if } A > 1 \text{ for } 0 \le t \le 1 \text{ ,} \\ > 0 \text{ , if } 0 < A < 1 \text{ for } 0 \le t < A \text{ ;} \end{cases}$$

$$F_{y'}[t+B-1] = -2(t+B-1) \begin{cases} < 0 \text{ , if } B > 1 \text{ for } 0 \le t \le 1 \text{ ,} \\ < 0 \text{ , if } 0 < B < 1 \text{ for } 1-B < t \le 1 \text{ ;} \end{cases}$$

$$F_{y'}[B+1-t] = 2(B+1-t) \begin{cases} < 0 \text{ , if } B < -1 \text{ for } 0 \le t \le 1 \text{ ,} \\ < 0 \text{ , if } -1 < B < 0 \text{ for } B+1 < t \le 1 \text{ .} \end{cases}$$

Consider first the case when either u_L or u_R is globally stable.

(i) $A < -1$. Since

$$\int_{\eta}^{u_L(1)=1+A} (-s)ds \not< 0 \quad \text{for } B \le \eta < 1+A \quad (\text{if } B < 1+A)$$

and

$$\int_{1+A}^{\eta} (-s)ds > 0 \quad \text{for } 1+A < \eta \le B \quad (\text{if } B > 1+A)$$

<u>provided</u> $|B| < -(1+A)$, we see that u_L does not support a $\cap$-boundary layer at $t = 1$, but that it does support a $\cup$-layer at $t = 1$ (if $B \le 0$) or an S-layer (if $B > 0$) there.

(ii) $B > 1$. Clearly

$$\int_{u_R(0)=B-1}^{\eta} (-s)\,ds \not> 0 \quad \text{for} \quad B-1 < \eta \leq A \ (\text{if} \ \ A > B-1), \text{ that is,}$$

u_R cannot support a U-layer at $t = 0$; however,

$$\int_{\eta}^{B-1} (-s)\,ds < 0 \quad \text{for} \quad A \leq \eta < B-1 \quad (\text{if} \ \ A < B-1)$$

<u>provided</u> $|A| < B-1$, that is, $-(B-1) < A < B-1$. Thus u_R does support a $\cap$-boundary layer (if $A \geq 0$) at $t = 0$ or a S-layer (if $A < 0$) there.

Consider now the functional

$$J[t] = \int_{t+A}^{t+B-1} (-s)\,ds \quad ;$$

it follows that $J[t^*] = 0$ only at $t^* = \frac{1}{2}(1-B-A)$ and that $J'[t] < 0$ for $0 \leq t \leq 1$. By the theory of Section 6 we know that if neither u_L nor u_R supports an S-layer at $t = 1$ or $t = 0$ then t^* belongs to $(0,1)$, that is, u_L and u_R support an S-shock layer at t^* . This occurs for $-A-1 < B < -A+1$.

Suppose next that u_L is globally stable $(A < -1)$ but that u_R is only stable on $[t_R,1]$ for $t_R = 1-B$ in $(0,1)$. If $B > -(A+1)$ then u_L cannot support an S-boundary layer at $t = 1$ and we know (cf. Section 6) that u_L and u_R support an S-shock layer at $t^* = \frac{1}{2}(1-B-A)$ in $(t_R,1)$. Similarly if u_L is only stable on $[0,t_L]$ for $t_L = -A(-1 < A < 0)$ and if u_R is globally stable $(B > 1)$ then u_L and u_R support an S-shock at t^* in $(0,t_L)$ provided $-A > B-1$ (that is, u_R cannot support an S-layer at $t = 0$) .

Now suppose that u_L and u_R are both stable only on $[0,t_L]$ and

$[t_R,1]$, respectively, that is, $-1 < A < 0$ and $0 < B < 1$. If $-A > 1-B$ then $t_L > t_R$ (the domains of stability overlap) and we know again from Section 6 that u_L and u_R support an S-shock layer at $t^* = \frac{1}{2}(1-B-A)$ in (t_R,t_L) . Next if $-A = 1-B$ then $u_L \equiv u_R$; indeed, $y(t,\epsilon) = t+A = t+B-1$ is an exact solution of (E5). Finally if $-A < 1-B$ the domains of stability do not overlap, that is, $u_L(t_L) = u_R(t_R) = 0$. However $u_s \equiv 0$ is $(I)_0$-stable and so by Theorem 6.4 (E5) has a solution $y = y(t,\epsilon)$ such that

$$\lim_{\epsilon\to 0^+} y(t,\epsilon) = \begin{cases} t+A\ , & 0 \le t \le -A\ , \\ 0\ , & -A \le t \le 1-B\ , \\ t+B-1, & 1-B \le t \le 1\ . \end{cases}$$

The same analysis just given for u_L,u_R applies analogously to $\tilde{u}_L,\tilde{u}_R$. We only summarize these results.

(i) If $A > 1$ $\tilde{u}_L$ is globally stable and $\tilde{u}_L$ cannot support a U-boundary layer at $t = 1$; however, if $B < A-1$ $\tilde{u}_L$ supports a ∩-boundary layer at $t = 1$ (if $B \ge 0$) or a Z-layer at $t = 1$ (if $-(A-1) < B < 0$) ;

(ii) If $B < -1$ $\tilde{u}_R$ is globally stable and $\tilde{u}_R$ cannot support a ∩-boundary layer at $t = 0$; however, if $A > B+1$ $\tilde{u}_R$ supports a U-boundary layer at $t = 0$ (if $A \le 0$) or a Z-layer at $t = 0$ (if $0 < A < -(B+1)$) ;

(iii) If $A > 1$ and $B < -1$ $\tilde{u}_L$ and $\tilde{u}_R$ support a Z-shock layer at $\tilde{t} = \frac{1}{2}(A+B+1)$ in $(0,1)$ if $-A-1 < B < -A+1$, that is, if $\tilde{u}_L$ cannot support a Z-layer at $t = 1$ and $\tilde{u}_R$ cannot support a Z-layer at $t = 0$; if $A > 1$ ($\tilde{u}_L$ globally stable) and $-1 < B < 0$

($\tilde{u}_R$ stable on $[B+1,1]$) and if $B < -(A-1)$ (that is, $\tilde{u}_L$ cannot support a Z-layer at $t = 1$) then $\tilde{u}_L$ and $\tilde{u}_R$ support a Z-shock at $\tilde{t}$ in $(B+1,1)$; if $B < -1$ ($\tilde{u}_R$ globally stable) and $0 < A < 1$ ($\tilde{u}_L$ stable on $[0,A]$) and if $A > -(B+1)$ (that is, $\tilde{u}_R$ cannot support a Z-layer at $t = 0$) then $\tilde{u}_L$ and $\tilde{u}_R$ support a Z-shock at $\tilde{t}$ in $(0,A)$.

(iv) If $0 < A < 1$ and $-1 < B < 0$ there are three subcases. If $A > B+1$ then the domains of stability of $\tilde{u}_L$ and $\tilde{u}_R$ overlap and these functions support a Z-shock at $\tilde{t} = \frac{1}{2}(A+B+1)$ in $(B+1,A)$. If $A = B+1$ then $\tilde{u}_L \equiv \tilde{u}_R$ and $y(t,\epsilon) = A-t = B+1-t$ is an exact solution of (E5) . Finally if $A < B+1$ $\tilde{u}_L(A) = 0$ and $\tilde{u}_R(B+1) = 0$ and so (E5) has a solution $y = y(t,\epsilon)$ such that

$$\lim_{\epsilon\to 0^+} y(t,\epsilon) = \begin{cases} A-t , & 0 \le t \le A , \\ 0 , & A \le t \le B+1 , \\ B+1-t , & B+1 \le t \le 1 . \end{cases}$$

Consider now the functions $u_L(t) = t+A$ and $\tilde{u}_R(t) = B+1-t$, and suppose first that each is globally stable, that is, $A < -1$ and $B < -1$. For this range of A and B $u_L, \tilde{u}_R$ are both negative and since the coefficient of y'^2 in (E5) is $-y$ we know (cf. Section 3) that u_L and $\tilde{u}_R$ can support only U- , S- or Z-boundary layers. Suppose however that $u_L(1) = 1+A > B$ and $\tilde{u}_R(0) = B+1 > A$, then we know (cf. Section 6) that u_L intersects $\tilde{u}_R$ at the point $t_0 = \frac{1}{2}(B-A+1)$ in $(0,1)$ and consequently, (E5) has a solution $y = y(t,\epsilon)$ such that

$$(\sim) \qquad \lim_{\epsilon\to 0^+} y(t,\epsilon) = \begin{cases} t+A , & 0 \le t \le t_0 , \\ B+1-t, & t_0 \le t \le 1 . \end{cases}$$

Similar behavior also occurs for $A < -1$ and $-1 < B < 0$ provided $-1 < B < A+1$, and for $B < -1$ and $-1 < A < 0$ provided $-1 < A < B+1$, that is, ($\sim$) is valid with t_0 in $(B+1,1)$ or $(0,-A)$. Suppose finally that $-1 < A$, $B < 0$. If $-A > B+1$ then the domains of stability of u_L and $\tilde{u}_R$ overlap and so ($\sim$) is valid at t_0 in $(B+1,-A)$. Next if $-A = B+1$ then $u_L(-A) = \tilde{u}_R(B+1) = 0$, that is, u_L and $\tilde{u}_R$ intersect at a point on the singular arc $u_s \equiv 0$; see figure 28.

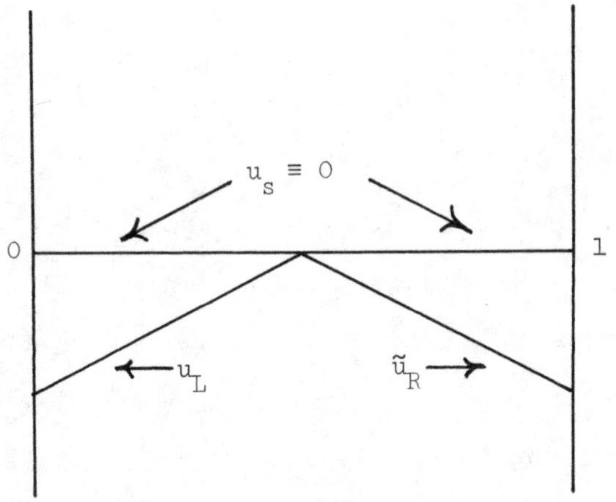

Fig. 28

Modified Haber-Levinson Crossing of u_L and $\tilde{u}_R$.

Since u_s is stable $(\sim)$ is valid here too. The last case is $-A < B+1$. Here u_L intersects u_s at $t = -A$ and $\tilde{u}_R$ intersects u_s at $t = B+1$. The stability of u_s implies that (E5) has a solution $y = y(t,\epsilon)$ such that

$$\lim_{\epsilon\to 0^+} y(t,\epsilon) = \begin{cases} t+A, & 0 \le t \le -A, \\ 0, & -A \le t \le B+1, \\ B+1-t, & B+1 \le t \le 1. \end{cases}$$

Our discussion of (E5) will now be complete if we consider finally the interaction of $\tilde{u}_L(t) = A-t$ and $u_R(t) = t+B-1$. Proceeding as in the case of u_L, $\tilde{u}_R$ we can show the following:

(i) If $A > 1$ and $B > 1$ there is a solution $y = y(t,\epsilon)$ of (E5) such that

$$(\approx) \qquad \lim_{\epsilon\to 0^+} y(t,\epsilon) = \begin{cases} A-t, & 0 \le t \le \tilde{t}_0, \\ t+B-1, & \tilde{t}_0 \le t \le 1, \end{cases}$$

where $\tilde{t}_0 = \frac{1}{2}(A-B+1)$ provided $A-1 < B < A+1$, that is, $\tilde{u}_L(1) < B$ and $u_R(0) < A$.

(ii) If $A > 1$ and $0 < B < 1$ (or $B > 1$ and $0 < A < 1$) then $(\approx)$ obtains with $\tilde{t}_0$ in $(1-B,1)$ (or $(0,A)$) provided $A-1 < B < 1$ (or $B-1 < A < 1$).

(iii) If $0 < A$, $B < 1$ then $(\approx)$ obtains if $A > 1 - B$ with $\tilde{t}_0$ in $(1-B,A)$. If $A = 1-B$, $\tilde{u}_L(A) = 0$ and $u_R(1-B) = 0$ but $(\approx)$ obtains because u_s is stable. Finally if $A < B - 1$ (E5) has a solution $y = y(t,\epsilon)$ such that

$$\lim_{\epsilon \to 0^+} y(t,\epsilon) = \begin{cases} A-t \ , & 0 \le t \le A \ , \\ 0 & A \le t \le 1-B \ , \\ t+B-1 \ , & 1-B \le t \le 1 \ . \end{cases}$$

We summarize our results in the form of a boundary value portrait; see figure 29.

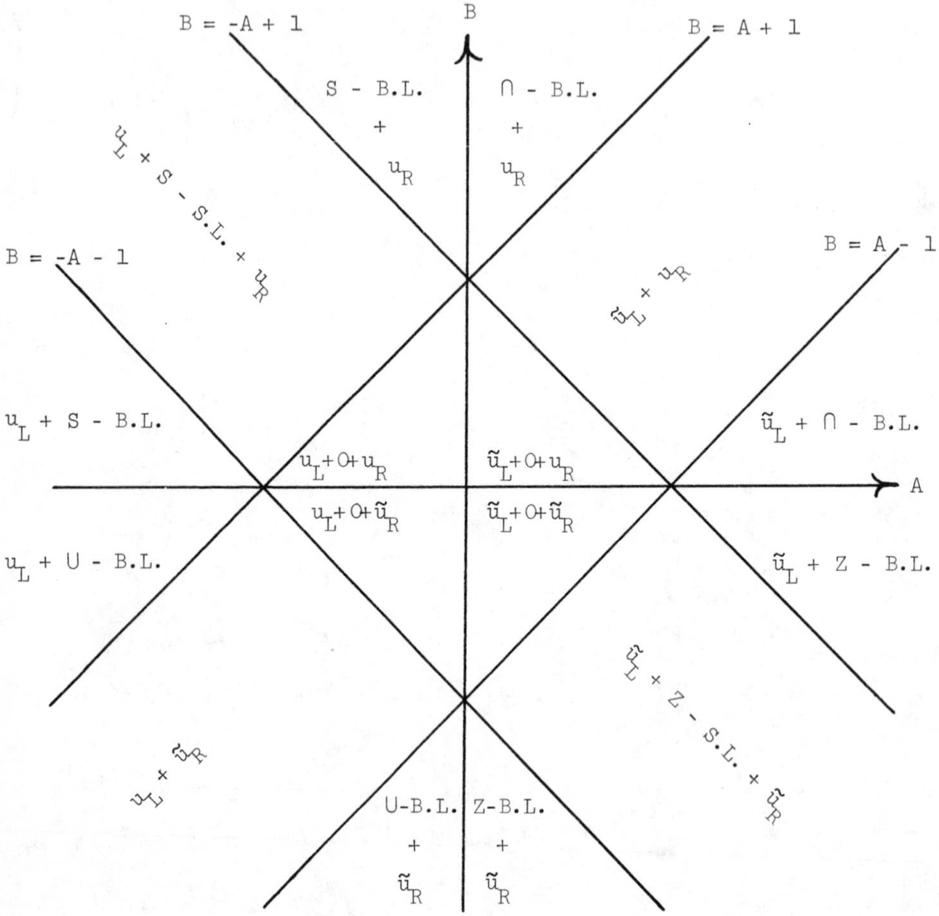

Fig. 29

Boundary Value Portrait of (E5) .

Consider next the problem

$$\epsilon y'' = yy'^2 - y = F(t,y,y') , \quad 0 < t < 1 ,$$

(E6)

$$y(0,\epsilon) = A, y(1,\epsilon) = B .$$

The reduced equation has the solutions $u_L(t) = t+A$, $u_R(t) = t+B-1$, $\tilde{u}_L(t) = A-t$, $\tilde{u}_R(t) = B+1-t$, $u_I(t) = t+C$, $\tilde{u}_I(t) = C-t$ and finally the singular solution $u_s \equiv 0$. Here C is any real constant. To check for stability we evaluate $F_{y'}$ and F_y along each of these functions:

$$F_{y'}[t+A] = 2(t+A) \begin{cases} > 0 , & \text{if } A > 0 \text{ for } 0 \le t \le 1 , \\ < 0 , & \text{if } A < 0 \text{ for } t = 0 ; \end{cases}$$

$$F_{y'}[t+B-1] = 2(t+B-1) \begin{cases} < 0 , & \text{if } B < 0 \text{ for } 0 \le t \le 1 , \\ > 0 , & \text{if } B > 0 \text{ for } t = 1 ; \end{cases}$$

$$F_{y'}[A-t] = -2t(A-t) \begin{cases} > 0 , & \text{if } A < 0 \text{ for } 0 \le t \le 1 , \\ < 0 , & \text{if } A > 0 \text{ for } t = 0 ; \end{cases}$$

$$F_{y'}[B+1-t] = -2(B+1-t) \begin{cases} < 0 , & \text{if } B > 0 \text{ for } 0 \le t \le 1 , \\ > 0 , & \text{if } B < 0 \text{ for } t = 1 ; \end{cases}$$

$$F_{y'}[t+C] = 2(t+C) \begin{cases} < 0 , & \text{if } C < 0 \text{ for } 0 \le t < \delta , \\ > 0 , & \text{if } C > -1 \text{ for } 1-\delta < t \le 1 ; \end{cases}$$

$$F_{y'}[C-t] = -2(C-t) \begin{cases} < 0 , & \text{if } C > 0 \text{ for } 0 \le t < \delta , \\ > 0 , & \text{if } C < 1 \text{ for } 1-\delta < t \le 1 ; \end{cases}$$

$$F_y[0] = -1 .$$

Thus u_L is (globally) stable if $A \geq 0$ and unstable if $A < 0$; however, u_L is stable with respect to boundary layer behavior at $t = 1$ if $-1 < A < 0$. Similarly u_R is (globally) stable if $B \leq 0$ and unstable if $B > 0$; however, u_R is stable with respect to boundary layer behavior at $t = 0$ if $0 < B < 1$. On the other hand $\tilde{u}_L$ is (globally) stable if $A \leq 0$ and unstable if $A > 0$; however, $\tilde{u}_L$ is stable with respect to layer behavior at $t = 1$ if $0 < A < 1$. And $\tilde{u}_R$ is (globally) stable if $B \geq 0$, unstable if $B < 0$ and stable with respect to boundary layer behavior at $t = 0$ if $-1 < B < 0$. Finally u_s is unstable. Thus the stability properties of these reduced solutions are markedly different from those of the problem (E5). The geometric properties of solutions of this problem are also different from those of (E5). For instance solutions of (E6) can display no shock layer behavior and all boundary layers are either $\cup$- or $\cap$-layers (cf. Section 3). With this in mind we consider now the asymptotic behavior of solutions of (E6), noting that for each $\epsilon > 0$ a solution $y = y(t,\epsilon)$ satisfying

$$t + \min\{A, B-1\} \leq y(t,\epsilon) \leq t + \max\{A, B-1\}$$

exists for all A and B.

If $A \geq 0$ then u_L is globally stable and clearly u_L supports a $\cup$-boundary layer at $t = 1$ for $B > 1+A$. If $B < 1+A$ there is no layer behavior at $t = 1$ because the coefficient of y'^2 in (E6) is positive for such A. Next if $A \leq 0$ $\tilde{u}_L$ is globally stable and $\tilde{u}_L$ supports only a $\cap$-layer at $t = 1$ for $B < A - 1$.

Similarly for $B \le 0$ u_R is globally stable and supports only a $\cap$-boundary layer at $t = 0$ if $B - 1 > A$. And for $B \ge 0$ $\tilde{u}_R$ is globally stable and supports a $\cup$-layer at $t = 0$ if $B+1 < A$.

Consider now the pairs $\{u_L, \tilde{u}_R\}$ and $\{\tilde{u}_L, u_R\}$. If $A, B > 0$ assume that $u_L(1) > B$ and $\tilde{u}_R(0) > A$, that is, $A-1 < B < A+1$.

From Section 6 we know that u_L intersects $\tilde{u}_R$ at the point $t_0 = \frac{1}{2}(B-A+1)$ in $(0,1)$ and that (E6) has a solution $y = y(t,\epsilon)$ such that

$$\lim_{\epsilon \to 0^+} y(t,\epsilon) = \begin{cases} t+A , & 0 \le t \le t_0 , \\ B+1-t , & t_0 \le t \le 1 . \end{cases}$$

Similarly if $A, B < 0$ assume that $\tilde{u}_L(1) < B$ and $u_R(0) < A$, that is, $A-1 < B < A+1$. We know that $\tilde{u}_L$ intersects u_R at $\tilde{t}_0 = \frac{1}{2}(A-B+1)$ in $(0,1)$ and therefore (E6) has a solution $y = y(t,\epsilon)$ such that

$$\lim_{\epsilon \to 0^+} y(t,\epsilon) = \begin{cases} A-t , & 0 \le t \le \tilde{t}_0 , \\ t+B-1 , & \tilde{t}_0 \le t \le 1 . \end{cases}$$

Suppose now that $A < 0$ and $B > 0$. (The reflected case $A > 0$ and $B < 0$ is handled similarly using $\tilde{u}_L$ and $\tilde{u}_R$ in place of u_L and u_R.) If $A < -1$ and $B > 1$ then by Nagumo's theorem (E6) has a solution $y = y(t,\epsilon)$ such that for $0 \le t \le 1$

$$t + A \le y(t,\epsilon) \le t + B - 1 .$$

However u_L and u_R are both unstable for such A and B. It follows that for a constant C in $(-1,0)$ (E6) has a solution $y = y(t,\epsilon)$ such that for $0 < t < 1$

$$\lim_{\epsilon\to 0^+} y(t,\epsilon) = t + C \quad .$$

The problem is now to find C . We begin by making the change of dependent variable $w = y-t$ which transforms (E6) into

$$\epsilon w'' = (w+t)w'^2 + 2(w+t)w' \; , \quad 0 < t < 1 \; ,$$

$$w(0,\epsilon) = A \; , \; w(1,\epsilon) = B-1 \quad .$$

Since $A < 0$ and $B-1 > 0$ $w'(t,\epsilon) > 0$ for $0 \le t \le 1$ and so we can rewrite the w-equation as

$$\epsilon \frac{w''}{w'} = (w+t)w' + 2(w+t) \; , \text{ that is,}$$

$$\epsilon \frac{d}{dt}(\ln w'(t,\epsilon)) = \frac{1}{2}\frac{d}{dt}(w^2) + tw' + 2(w+t) \quad .$$

Integrating both sides of this equation from 0 to 1 , using the boundary conditions and integrating the resulting integrals by parts we have that

$$(\sim) \qquad \epsilon \ln(w'(1,\epsilon)) - \epsilon \ln(w'(0,\epsilon)) = \frac{1}{2}(B-1)^2 - \frac{1}{2}A^2 + B + \int_0^1 w(t,\epsilon)dt \; .$$

Now by the estimates of Vishik and Liusternik [34] (cf. also [32; Chap. 2)]

$$w'(0,\epsilon) = \exp[-\epsilon^{-1}\int_A^C s\, ds] \quad \text{and}$$

$$w'(1,\epsilon) = \exp[\epsilon^{-1}\int_C^{B-1} (s+1)ds] \quad ,$$

that is,

$$\epsilon \ln(w'(0,\epsilon)) = -\int_A^C s\, ds = \frac{1}{2}A^2 - \frac{1}{2}C^2 \quad \text{and}$$

$$\epsilon \ln(w'(1,\epsilon)) = \int_C^{B-1} (s+1)ds = \frac{1}{2}(B-1)^2 + (B-1) - \frac{1}{2}C^2 - C \quad .$$

Here C is the constant we are trying to determine, namely, $\lim_{\epsilon\to 0^+} w(t,\epsilon) = C$ for $0 < t < 1$. Substituting into $(\sim)$ and simplifying gives

$$-1 - C = \int_0^1 w(t,\epsilon)dt \quad .$$

Letting $\epsilon \to 0+$ in this expression, we obtain finally $C = -\frac{1}{2}$ since $\lim_{\epsilon\to 0^+} \int_0^1 w(t,\epsilon)dt = C$. In terms of (E6) this means that for $B > 1$ and $A < -1$ the solution $y = y(t,\epsilon)$ satisfies

$$\lim_{\epsilon\to 0^+} y(t,\epsilon) = t - \frac{1}{2} \quad \text{for} \quad 0 < t < 1 \quad ,$$

that is, $u_I(t) = t - \frac{1}{2}$ supports a $\cap$-layer at $t = 0$ and $\cup$-layer at $t = 1$. The remaining subcases are handled similarly, namely solutions $y = y(t,\epsilon)$ of (E6) exist and satisfy $\lim_{\epsilon\to 0^+} y(t,\epsilon) = t+A,\ t+B-1$, and/or $t = \frac{1}{2}$ depending on the relative position of A and B. We note however one important difference between this case and the corresponding one of (E4). Suppose that $-\frac{1}{2} < A < 0$, $0 < B < 1$ and $B - 1 < A$. Nagumo's theorem guarantees the existence of at least one solution $y(t,\epsilon)$ of (E6) such that for $0 \leq t \leq 1$

$$t + B - 1 \leq y(t,\epsilon) \leq t + A \quad .$$

Although u_L is stable with respect to boundary layer behavior at $t = 1$ $u_L(1) > B$ and we know that u_L cannot support a $\cap$-layer there. Similarly u_R is stable with respect to boundary layer behavior at $t = 0$ but $u_R(0) < A$ and u_R cannot support a $\cup$-layer there. Consider now u_L and observe that $u_L(1) > B$ implies that u_L

intersects $\tilde{u}_R$ at $t_0 = \frac{1}{2}(B-A+1)$ in $(-A,1)$. Arguing as in [18] we can show that (E6) has a solution $y_1 = y_1(t,\epsilon)$ such that

$$\lim_{\epsilon\to 0^+} y_1(t,\epsilon) = \begin{cases} t+A\ , & 0 \le t \le t_0\ , \\ B+1-t\ , & t_0 \le t \le 1\ . \end{cases}$$

Similarly since $u_R(0) < A$ it is clear that u_R intersects $\tilde{u}_L$ at $\tilde{t}_0 = \frac{1}{2}(A-B+1)$ in $(0,1-B)$ and so (E6) has another solution $y_2 = y_2(t,\epsilon)$ such that

$$\lim_{\epsilon\to 0^+} y_2(t,\epsilon) = \begin{cases} A-t\ , & 0 \le t \le \tilde{t}_0\ , \\ t+B-1\ , & \tilde{t}_0 \le t \le 1\ ; \end{cases}$$

see figures 30 and 31.

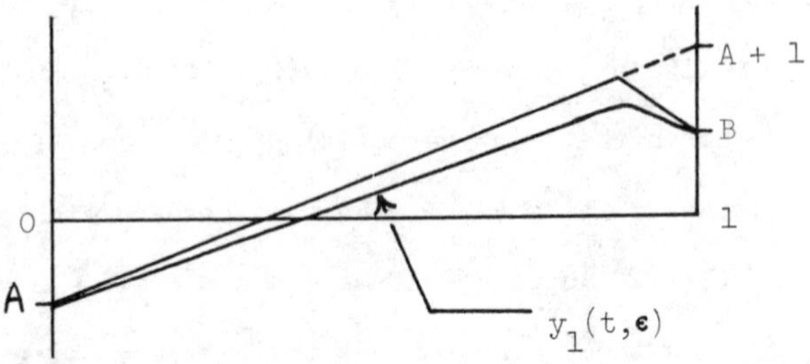

Fig. 30

Solution of (E6) generated by the crossing of u_L and $\tilde{u}_R$.

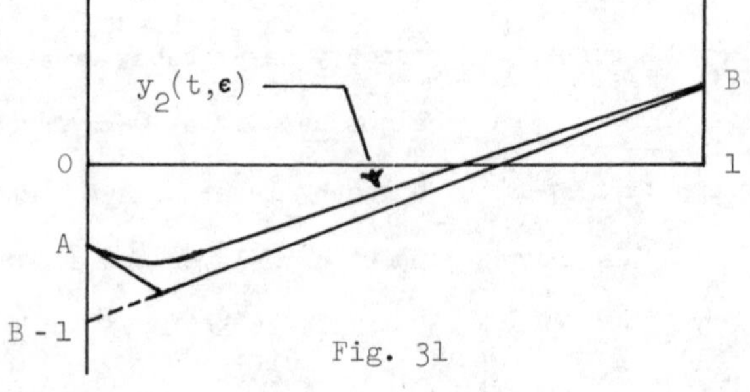

Fig. 31

Solution of (E6) generated by the crossing of $\tilde{u}_L$ and u_R.

Consider now the related problems

$$\epsilon y'' = y + yy'^2 = y(1+y'^2) , \quad 0 < t < 1 ,$$

(E6)'

$$y(0,\epsilon) = A, y(1,\epsilon) = B ,$$

and

$$\epsilon y'' = -y - yy'^2 = -y(1+y'^2) , \quad 0 < t < 1 ,$$

(E6)"

$$y(0,\epsilon) = A, y(1,\epsilon) = B .$$

It was shown in [16] that (E6)' has a solution $y = y(t,\epsilon)$ for all $\epsilon > 0$ and all A, B such that

$$\lim_{\epsilon \to 0^+} y(t,\epsilon) = 0 \quad \text{for} \quad 0 < t < 1$$

(with convergence at $t = 0$ $(t=1)$ if $A = 0$ $(B = 0)$) . Our previous considerations make such a result reasonable because $u_s \equiv 0$ is (I_0)-stable and $(1+y'^2)$ is bounded away from zero.

As for (E6)" the situation is quite different because $u_s \equiv 0$ (the only real reduced solution) is unstable. It is not hard to see that (E6)" has no solutions of bounded t-variation for ϵ small if $A \neq 0$ or $B \neq 0$.

We consider finally an example of (P_3) which is not entirely amenable to our techniques. The problem is

$$\epsilon y'' = -yy'^2 + yy' , \quad 0 < t < 1 ,$$

(E7)

$$y(0,\epsilon) = A , \quad y(1,\epsilon) = B ,$$

and let us assume that $A > 0$ and $B < 0$. Clearly by Nagumo's theorem (E7) has a solution $y = y(t,\epsilon)$ such that for $0 \le t \le 1$

$$B \le y(t,\epsilon) \le A \quad .$$

Since $A > 0$ $u_L \equiv A$ is globally stable and since $B < 0$ $u_R \equiv B$ is also globally stable.

Consider first the occurrence of boundary layer behavior at $t = 0$. It follows that

$$\int_{\xi}^{u_R(0)=B} (-s)ds > 0 \quad \text{for} \quad B < \xi \le A \quad \underline{\text{provided}} \quad |A| < -B \quad ,$$

that is, if $B < A < -B$ u_R supports a U-boundary layer (if $A \le 0$) or a Z-boundary layer (if $A > 0$) at $t = 0$. Similarly at $t = 1$

$$\int_{\eta}^{u_L(1)=A} (-s)ds < 0 \quad \text{for} \quad B \le \eta < A \quad \underline{\text{provided}} \quad |B| < A \quad ,$$

that is, if $-A < B < A$ u_L supports a ∩-boundary layer (if $B \ge 0$) or a Z-boundary layer (if $B < 0$) at $t = 1$. Neither u_L nor u_R can support a Z-layer if $B \le -A$ <u>and</u> $A \ge -B$, namely if $A = -B$. For such A and B

$$J[t] = \int_B^A (-s)ds \equiv 0 \quad \text{and so} \quad J'[t] \equiv 0$$

thus making our previous shock layer theory inapplicable. We know however that u_L and u_R support a Z-shock layer in $(0,1)$. The problem is then to find the t^* in $(0,1)$ such that

$$(*) \qquad \lim_{\epsilon\to 0^+} y(t,\epsilon) = \begin{cases} A \,, & 0 \le t < t^* \,, \\ B \,, & t^* < t \le 1 \,. \end{cases}$$

We argue as follows. From the form of (E7) it is clear that

$y'(t,\epsilon) < 0$ for $0 \leq t \leq 1$ and so the y-equation can be rewritten as

$$\epsilon \frac{y''}{y'} = - yy' + y \ , \text{ that is, } \epsilon \frac{d}{dt}(\ell n|y'|) = -\frac{1}{2}\frac{d}{dt}(y^2) + y \ .$$

Now we integrate both sides of this equation from 0 to t^* and from t^* to 1 and use the boundary conditions to arrive at the two equations:

$$\epsilon \ell n|y'(t^*,\epsilon)| - \epsilon \ell n|y'(0,\epsilon)| = -\frac{1}{2}(y(t^*)^2 - A^2) + \int_0^{t^*} y(t,\epsilon)dt$$

and

$$\epsilon \ell n|y'(1,\epsilon)| - \epsilon \ell n|y'(t^*,\epsilon)| = -\frac{1}{2}(B^2 - y(t^*)^2) + \int_{t^*}^1 y(t,\epsilon)dt \ .$$

Adding these equations gives finally

$$\epsilon \ell n|y'(1,\epsilon)| - \epsilon \ell n|y'(0,\epsilon)| = \frac{1}{2}(A^2 - B^2) + (\int_0^{t^*} + \int_{t^*}^1)y(t,\epsilon)dt \ ,$$

that is,

$$(\sim) \qquad (\int_0^{t^*} + \int_{t^*}^1)y(t,\epsilon)dt = 0$$

because $y'(0,\epsilon) = y'(1,\epsilon)$ and $A = -B$.

By $(*)$ $\lim_{\epsilon\to 0+} y(t,\epsilon) = A$ for $0 \leq t < t^*$ and

$\lim_{\epsilon\to 0+} y(t,\epsilon) = B$ for $t^* < t \leq 1$; consequently,

$$\lim_{\epsilon\to 0+} \int_0^{t^*} y(t,\epsilon)dt = At^* \text{ and}$$

$$\lim_{\epsilon\to 0+} \int_{t^*}^1 y(t,\epsilon)dt = B(1-t^*) = -A(1-t^*) \ .$$

Letting $\epsilon \to 0+$ in $(\sim)$ we have that $t^* = \frac{1}{2}$. In summary, for $A = -B$ (E7) has a solution $y = y(t,\epsilon)$ such that

$$\lim_{\epsilon\to 0+} y(t,\epsilon) = \begin{cases} A \ , & 0 \leq t < \frac{1}{2} \ , \\ B \ , & \frac{1}{2} < t \leq 1 \ . \end{cases}$$

A discussion of the remaining cases can be effected without difficulty.

References

1. Yu. P. Boglaev, The Two-Point Problem for a Class of Ordinary Differential Equations with a Small Parameter Coefficient of the Derivative, USSR Comp. Math. Phys. 10 (1970), 191-204.

2. J.M. Burgers, A Mathematical Model Illustrating the Theory of Turbulence, in "Advances in Applied Mechanics", vol. 1, ed. by v. Mises and v. Karman, pp. 171-199, 1948.

3. G.F. Carrier and C.E. Pearson, Ordinary Differential Equations, Ginn/Blaisdell, Waltham, Mass., 1968.

4. K.W. Chang, On Coddington and Levinson's Results for a Nonlinear Boundary Value Problem Involving a Small Parameter, Rend. Acad. Lincei 54 (1973), 356-363.

5. E.A. Coddington and N. Levinson, A Boundary Value Problem for a Non-Linear Differential Equation with a Small Parameter, Proc. Amer. Math. Soc. 3 (1952), 73-81.

6. J.D. Cole, Perturbation Methods in Applied Mathematics, Ginn/Blaisdell, Waltham, Mass., 1968.

7. F.W. Dorr, S.V. Parter and L.F. Shampine, Application of the Maximum Principle to Singular Perturbation Problems, SIAM Rev. 15 (1973), 43-88.

8. A. Erdélyi, On a Nonlinear Boundary Value Problem Involving a Small Parameter, J. Austral. Math. Soc. 2 (1962), 425-439.

9. P.C. Fife, Semilinear Elliptic Boundary Value Problems with Small Parameters, Arch. Rational Mech. Anal. 52 (1973), 205-232.

10. ------------, Transition Layers in Singular Perturbation Problems, J. Diff. Eqns. 15 (1974), 77-105.

11. ------------, Two-Point Boundary Value Problems Admitting Interior Transition Layers, unpublished.

12. ------------, Boundary and Interior Transition Layer Phenomena for Pairs of Second-Order Differential Equations, J. Math. Anal. Appl. 54 (1976), 497-521.

13. J.E. Flaherty and R.E. O'Malley, Jr., The Numerical Solution of Boundary Value Problems for Stiff Differential Equations, Math. Comp. 31 (1977), 66-93.

14. S. Haber and N. Levinson, A Boundary Value Problem for a Singularly Perturbed Differential Equation, Proc. Amer. Math. Soc. 6 (1955), 866-872.

15. J.W. Heidel, A Nonlinear Second Order Boundary Value Problem, J. Math. Anal. Appl. 48 (1974),

16. F.A. Howes, A Class of Boundary Value Problems Whose Solutions Possess Angular Limiting Behavior, Rocky Mtn. J. Math. 6 (1976), 591-607.

17. ---------------, Singular Perturbations and Differential Inequalities, Memoirs Amer. Math. Soc. 168 (1976), 75 pp.

18. ---------------, The Asymptotic Solution of a Class of Singularly Perturbed Nonlinear Boundary Value Problems via Differential Inequalities, SIAM J. Math. Anal. 9 (1978).

19. ---------------, Singularly Perturbed Nonlinear Boundary Value Problems with Turning Points, II, ibid. 9 (1978).

20. ---------------, Singularly Perturbed Boundary Value Problems with Angular Limiting Solutions, Trans. Amer. Math. Soc., to appear.

21. ---------------, A Boundary Layer Theory for a Class of Linear and Nonlinear Boundary Value Problems, Rocky Mtn. J. Math. 7 (1977).

22. ---------------, Singularly Perturbed Nonlinear Boundary Value Problems Whose Reduced Equations have Singular Points, Studies in Appl. Math. 57 (1977), 135-180.

23. ---------------, Singularly Perturbed Superquadratic Boundary Value Problems, Nonlinear Anal., submitted for publication.

24. F.A. Howes and S.V. Parter, A Model Nonlinear Problem having a Continuous Locus of Singular Points, Studies in Appl. Math., to appear.

25. L.K. Jackson, Subfunctions and Second Order Ordinary Differential Inequalities, Adv. in Math. 2 (1968), 307 - 363.

26. J.D. Murray, On Burgers' Model Equations for Turbulence, J. Fluid Mech. 59 (1973), 263-279.

27. M. Nagumo, Über die Differentialgleichung $y'' = f(x,y,y')$, Proc. Phys. Math. Soc. Japan 19 (1937), 861-866.

28. R.E. O'Malley, Jr., On a Boundary Value Problem for a Nonlinear Differential Equation with a Small Parameter, SIAM J. Appl. Math. 17 (1969), 569-581.

29. ---------------, Introduction to Singular Perturbations, Academic Press, New York, 1974.

30. -------------------, Boundary Layer Methods for Ordinary Differential Equations with Small Coefficients Multiplying the Highest Derivatives, in Proc. Symp. on Constructive and Computational Methods for Diff'l and Integral Eqns., Springer Verlag Lecture Notes in Math., 430, 1974, pp. 363-389.

31. -------------------, Phase-Plane Solutions to Some Singular Perturbation Problems, J. Math. Anal. Appl. 54 (1976), 449-466.

32. A.B. Vasil'eva, Asymptotic Behavior of Solutions to Certain Problems Involving Nonlinear Differential Equations Containing a Small Parameter Multiplying the Highest Derivatives, Russian Math. Surveys 18 (1963), 13-84.

33. -------------------, Almost Discontinuous Solutions of a Conditionally Stable System with a Small Parameter Multiplying the Derivatives, Diff. Eqns. 8 (1972), 1204-1209.

34. M.I. Vishik and L.A. Liusternik, Initial Jump for Nonlinear Differential Equations Containing a Small Parameter, Sov. Math. Dokl. 1 (1960), 749-752.

35. W.R. Wasow, Singular Perturbations of Boundary Value Problems for Nonlinear Differential Equations of the Second-Order, Comm. Pure Appl. Math. 9 (1956), 93-113.

36. -------------------, Asymptotic Expansions for Ordinary Differential Equations, Interscience, New York, 1965.

37. -------------------, The Capriciousness of Singular Perturbations, Nieuw Arch. Wisk. 18 (1970), 190-210.

SCHOOL OF MATHEMATICS
UNIVERSITY OF MINNESOTA
MINNEAPOLIS, MINNESOTA 55455